STATION NORMAL

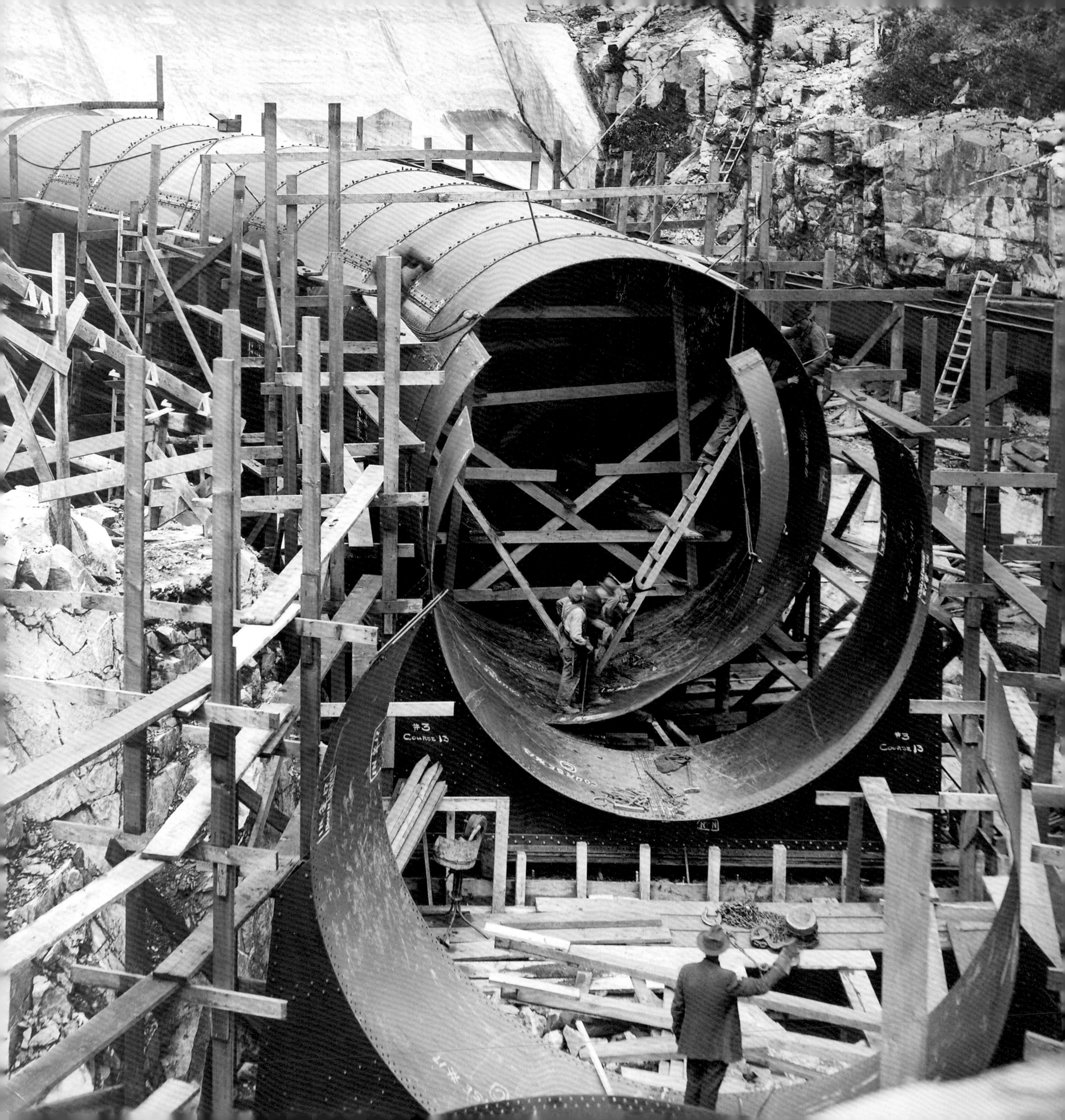
#3
COURSE 13
#3
COURSE 13

MEG STANLEY &
HUGH WILSON

STATION NORMAL

THE POWER OF THE STAVE RIVER

DOUGLAS & MCINTYRE
VANCOUVER/TORONTO

frontispiece:
Assembling penstock number 5, May 10, 1925. This is a familiar scene; one man is up the ladder, at work, and lots of people are watching. The diameter of the fifth penstock tapers from about six metres at its intake in the main dam to about four metres where it joins the turbine at the western end of the powerhouse.

01 02 03 04 05 5 4 3 2 1

Published by
Douglas & McIntyre Ltd.
2323 Quebec Street, Suite 201
Vancouver, British Columbia V5T 4S7

National Library of Canada Cataloguing in Publication Data
Stanley, Meg, 1962–
Station normal
Published in association with: B.C. Hydro.
Includes index.
ISBN 1-55054-932-4
1. Hydroelectric power plants—British Columbia—Stave Falls—History. 2. Stave Falls (B.C.)—History. I. Wilson, Hugh (Hugh Patrick). II. B.C. Hydro. III. Title.
HD9685.C33B7 2001 333.793'2'0971137 C2001-911529-6

Produced by
Commonwealth Historic Resource Management Limited
308 – 2233 Burrard Street
Vancouver, British Columbia V6J 3H9

Project management by Harold Kalman
Editing by John Eerkes-Medrano
Cover and text design by Peter Cocking
Front cover photo: BC Hydro, Stave Falls Records
Back cover photos, left to right: courtesy of Doug Noble and family; BC Hydro Information Centre B1092/97; courtesy of Kenneth and Vera Ruddick; courtesy of Dorothy and Len Winders; courtesy of Frances Fleming
Maps by Stuart Daniel/Starshell Maps
Printed and bound in Canada by Friesens

Care has been taken to trace the ownership of the illustrations and other materials used in the text. The publishers welcome information that will help them to correct any reference or credit in future editions of this book.

The publisher gratefully acknowledges the financial support of the Canada Council for the Arts, the British Columbia Ministry of Tourism, Small Business and Culture, and the Government of Canada through the Book Publishing Industry Development Program (BPIDP) for its publishing activities.

Contents

People and Power

THE STAVE RIVER, from its glacier source in the Pacific Ranges of the Coast Mountains to its junction with the Fraser River at Ruskin, is about one hundred kilometres long. For most of the 20th century, people have transformed the river's power into hydroelectric energy. Electrical power from the Stave helped the residents and industries of British Columbia's Lower Mainland and Fraser Valley to prosper during the early years of urbanization, and it has steadily contributed to the province's energy production since that time. Until the year 2000, this was accomplished with dams and a powerhouse first built more than ninety years ago.

If someone had told the builders of the Stave Falls dam and powerhouse that the two structures they worked so hard on would still be standing—let alone producing power—almost ninety years later, they would have been very impressed and proud. And perhaps a little surprised. When the power first started flowing from Stave Falls in 1912, the hydroelectric industry was still in its infancy. Relatively little theory existed on concepts like foundation uplift or reinforced concrete construction. The water records for the Stave watershed didn't go back many years. Earthquake preparedness? Hardly considered at all.

Yet, as the year 2000 approached, Stave Falls was still churning out useful, inexpensive electricity, much as it had done for most of the century. It did so with generators that had become historical artifacts and that were kept going by a select group of operators, mechanics, and electricians who passed on their knowledge to the next generation. Even as the machines were eclipsed again and again by superior technology—as spare parts became harder and then impossible to find—they kept doing the job.

At the end of each shift, each day, every month, year after year, staff at Stave Falls (as at other plants) would sign off in the logbook. If everything was fine, they would write "station normal," which meant all was well; any problems had been dealt with.

Station normal. These two words perfectly describe what Stave Falls was all about. In the most practical sense, the plant quietly and reliably did its job for most of the 20th century. At first it was one of the most important cogs in the power grid, and then, over time, it became a bit player. But there was also a normalcy in a human sense.

Many lives were lived at Stave Falls, which produced both people and power. Stave Falls was home to generations of employees, first with the Western Canada Power Company, then the BC Electric Company, and then BC Hydro. People lived at Stave Falls, were born there, grew up there, and sometimes died there. They built homes and community centres, barbershops and doctors' offices, tennis and badminton courts.

This book tells stories about Stave Falls and, to a lesser extent, about nearby Ruskin and Alouette. It is not a technical look at how power is produced, although that will be discussed. It is not a social history of life in a small town, although many unique aspects of that community shine through. Instead, the purpose of *Station Normal* is to pay homage to Stave Falls through the stories and memories of the people who knew it best. Here are stories of engineers, businessmen, financiers, labourers, operators, supervisors, husbands, wives, and children. Among these people were visionaries, foot soldiers, heroes, and a few rogues.

But Stave Falls is not merely a historical site. Although the original powerhouse was decommissioned in 2000 and was transformed into a public facility to tell many of these stories, power still flows from the Stave. And although only one caretaker now lives in the historic townsite and modern hydroelectric technologies require fewer workers to manage them, the dams and powerhouses are still all about people.

We hope you enjoy these stories and that they offer some insight into the operations at Stave Falls.

Six men on a power pole, fooling around at Stave Falls.

The mountain peaks at the northwestern end of Stave Lake are especially striking in this 1924 photograph. The two highest are Mount Robie Reid and Mount Judge Howay, locally known as the Snow Peaks.

CHAPTER ONE

The Beginnings: A Land of Marvellous Beauty

THE STAVE RIVER, its headwaters, natural reservoir, and valley were formed by the same powerful forces that define all the geology and geography of Canada: moving water, rock, and ice. Over tens of millions of years, these unstoppable forces shaped and reshaped the Canadian landscape, carving rivers, gorges, mountain ranges, and massive lakes and inland seas.

The underlying granite of the Stave area was formed during the Paleozoic and Mesozoic eras, starting about 600 million years ago and ending 100 million years ago. A series of glaciations between 24,000 and 11,000 years ago known as the Fraser Glaciation did not alter the fundamental mountain and valley landscape, but they did leave behind considerable debris that redirected the channel of the Stave River. E.E. Carpenter, who became BC Electric's chief engineer in 1925, succinctly described the impact of glaciation on the Stave:

[It] formed a series of morainal dams, the remnants of which, combined with occasional spurs of bedrock, form the barrier that holds the lake in place. These dams have been eroded away by stages and the present river valley, with its terraced flanks and occasional gorges, has been cut deeply into the stranded remnants of bedrock that withstood the grinding glacier but have proved a futile barrier to the gentler but persisting action of the stream.

By five thousand years ago, the area's geographical features had settled to become those that are familiar today. Civil engineer Robert F. Hayward, the man most publicly associated with the Stave Falls dam and power plant as they neared completion, wrote this description of the area in 1915:

The whole of the watershed is formed by the granite mountains of the Coast Range; the lower part, including half the lake and the Lower Stave River, which is twelve miles long, has a maximum altitude of about 4,000 feet and is heavily timbered; the upper part consists of rugged mountains rising high above the timber line, and carrying perpetual snow and small glaciers. Mount Baldy [now Mount Robie Reid], the principal peak of the watershed, standing near the head of the lake and rising to an altitude of over 6,000 feet when viewed from the river or lake on a fine day, forms one of the most beautiful pictures that can be found in British Columbia.

The upper Stave River's headwaters are in what is now the southern part of Garibaldi Provincial Park. Although these waters remained largely unexplored by government surveyors until the 1930s, First Nations people had established a number of trails through parts of the watershed over many hundreds of years. Later, prospectors criss-crossed the area in their typically determined way, with little regard for the challenges of topography. Nevertheless, the terrain surrounding the lake and upper Stave River remains very challenging. A local newspaper described the pre-dam lake in grandiloquent style in 1891:

Oxygen is everywhere, ozone between; birds are singing, the slopes are clothed and terraced with fir and vines; everywhere bloom and colour under the blue sky. The crashing sounds of the mountain torrent, as it leaps over the shelving rocks, and the gentle rippling of the hill-side brook, is everywhere to be heard. A few miles further, and then the lake. Bold and immense stands out in startling silhouette three snow-capped, ragged-edged mountains that burst upon the view; and lying calm and serene at their base, nestles this sheet of water, alive with trout. On the west, abrupt shores and frowning battlements of rock; on the north stupendous peaks with everlasting caps of ice and snow; on the east gentle slopes about four miles broad and ten long.

Stave Falls in about 1900. The river fell twenty-five metres from still water above the head of the falls to the pool at the foot of the falls.

The look of the lake, the river, and its shorelines changed as the man-made reservoir started to rise in 1911 and continues to rise and fall to this day.

Long before the dams were wedged between granite walls and crumbling banks, the lower Stave River ran a brief but tumultuous twenty-kilometre course from the foot of the lake to the Fraser River. From the lake, the Stave dropped smoothly for almost eleven kilometres, tumbling through one short rapid. About two and a half kilometres above the falls, the turbulence increased and then, suddenly, a plume of mist appeared where the water dropped precipitously over Stave Falls.

The falls were formed because of a small geological irregularity. On the east side of the Stave River valley, a low granitic spur (or knoll) extends from the base of a mountain, creating a distinct ridge across the river. Because granite hardly wears at all, the falls were created over this ledge. As the spur approaches the west bank it abruptly dips, and the gap is filled with glacial debris and gravel.

Before the dams were built, there were three channels at Stave Falls. The river flowed through the two western gaps. It was through the middle channel that

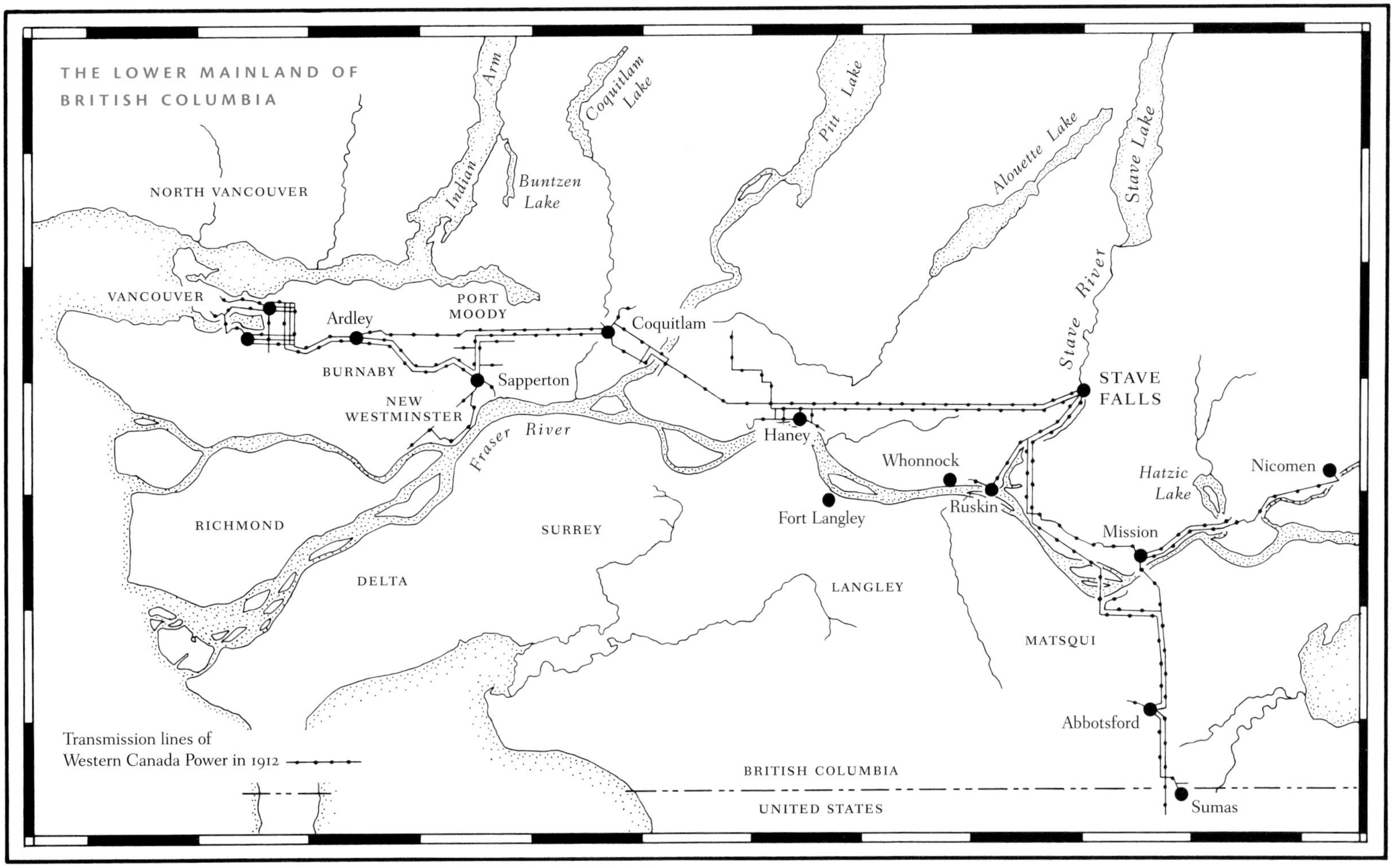

the most spectacular section of the falls plummeted. At one time most, if not all, of the river ran through the east channel, but at some point it became congested by successive log jams and beaver dams. This forced the flow of water to the west and the channel ran dry, inspiring the name "Blind Slough."

Below the falls were two beautiful wooded islands, where salmon had their spawning grounds on the river; the falls were impassable even by the powerful and persistent sockeye salmon. For another eight kilometres, the Stave flowed alternately as a series of rapids and as a placid, meandering river. Then, after rushing through a narrow granite gorge and flowing by more wooded islands, the river fanned out into the verdant and welcoming Fraser River lowlands.

The Stave is still a spectacular mix of dense woods, granite outcrops, towering peaks, and open terraces. But the area has been logged for well over one hundred years. Many of the slopes were clear-cut and are in various stages of regrowth; most of the terraces are underwater. For years, the lake was a watery minefield full of deadheads and snags because the reservoir was not completely cleared before it was flooded. Only when the lake is drawn down can the old river channel be seen.

THE "TIRELESS HUNTERS"

The Stave watershed has two parts. The more northerly area comprises most of the lake, feeder streams, and surrounding mountains; the terrain is steep and rugged. The more southerly part, from the lower part of Stave

Lake, the falls, canyons, and outlet, is more hospitable due to its glacial benches and flat areas.

Many descriptions of the area in the 19th and early 20th centuries included the words "inaccessible" and "uninhabited." The canyons of the lower Stave were usually described as forbidding barriers to anyone wishing to reach the falls or Stave Lake. In his 1915 description, Robert F. Hayward had this to say about the area:

Though the first settler entered the Stave Lake district over thirty years ago, access was so difficult that only a few families stayed on the land. To all practical purposes the valley of the Stave remained in its primitive state until the formation of the Stave Lake Power Company in 1899.

Although the upper Stave Lake area has always been too rugged for humans to settle, the Stave River was a permanent home for some of the Kwantlen people, and there were many routes to and through it. To the Kwantlen, the Stave was always "the source."

Snow-capped peaks dot the horizon looking east from Mount Baldy (now Mount Robie Reid) on the northwest shore of Stave Lake in 1944.

It is unlikely that humans appeared on the scene much earlier than ten thousand years ago, when the Fraser Glaciation was finally subsiding. Archaeological surveys of the lake bed, conducted by BC Hydro and the Kwantlen First Nation, found more than two thousand artifacts—primarily arrowheads, tools for dressing fish, and beads. In addition, pictographs were discovered on the shoreline of the Stave Reservoir. These discoveries make it clear that parts of the Stave watershed have been used for thousands of years.

The Kwantlen appear to have had three villages on the lower Stave: at the falls, near present-day Ruskin, and at the confluence with the Fraser River. Extended families and groupings of families lived in shed-roofed cedar longhouses made of logs and planks that were produced by using traditional tools such as stone chisels and adzes. The Kwantlen also established temporary sites in the watershed that were part of trade and hunting routes and that were used for spiritual purposes such as winter dancing.

These inhabitants of the Stave River area were part of the larger Kwantlen nation, whose territory stretched from present-day New Westminster to Xa:ytem (Hatzic) Rock, east of Mission. The Kwantlen in turn are part of the larger Stó:lō nation associated with the lower Fraser River. The Stó:lō people speak the Halkomelem language and were fishers, hunters, and traders. Kwantlen means "tireless hunter," and in the rugged territory of the Stave River and Stave Lake, that attribute would have been very useful.

Clarence Cheer, a Kwantlen who grew up in the village near the falls in the early 1900s, remembered a way of life that had been practised for millennia:

Some [residents] stayed all year round, and some of them didn't. I was raised in a longhouse . . . That's where everybody was raised.

We were taught to get up early in the morning. The people used to tell us about the water, that you would have to get down there and look at it. The water was still

Standing on solid ground close to the precipice at Stave Falls, about 1912.

asleep . . . And we would go down early in the morning and take our bath, come back . . . They didn't have the medicines and things we have today, and that was one reason they used to make us get tough to the cold. We didn't get sick as much as we do now . . .

In the mornings we didn't eat. We learned how to . . . prepare fish and meats and all that. We had to do that when we were young. We had to learn because we . . . had to look after ourselves and everything we had done. All our meats and hides—we had to . . . take off the hair and prepare it for buckskin and that was some of their trading material. We didn't throw the hides away—we had to know what to put on the hides so the hair would come off easy . . .

We were educated that we had to learn all [about] fishing, what parts to take and what parts not to take. Everything they done back there was trading . . . because we didn't know what money was. That's why they did so much trading, that was the money. They had to learn that the people from the Fraser River were trading with people over here because we don't get all that and we were trading buckskins. Later some of the smart young people would get pack horses and come over, and trade fish, and get more fish; it was a business.

Stave Lake was part of an overland and water route to the interior of the province. The people of the lower Fraser traded extensively with people both in the inte-

Artist Brandon Gabriel's recent drawing of Kwantlen life portrays a father and son watching Kwantlen people fish for sturgeon at their village near present-day Maple Ridge.

rior and on the coast. The hundreds of traditional trade goods included salmon (prepared in various ways), sturgeon oil, cedar baskets, canoes, goat and dog wool, buckskin, huckleberries, saskatoon berries, dried shellfish, wild potatoes, and seal skins.

Charlie Miller, who was born in England, spent almost his entire life near the Stave River and had a close relationship with the Kwantlen people there. He wrote about one of the trade routes in his book *Valley of the Stave*:

A well-defined trail progressed eastward along the high ground north of what is now Haney, along the south bank of the Alouette, then named Lillooet, to the foot of the lake. Here, a natural low-water crossing enabled them to proceed overland again, to higher land overlooking Stave Lake, then to a point on the lake now known as Morgan, formerly Alice Lake, and then down the trail beside it to a creek that ended at Stave Lake. Here, [at the beginning of the 20th century] much evidence of cedar dug-out canoe building could be observed. Albert Kearsley, who logged the Stave Lake area for many years, told me he had found the waterlogged remains of sixteen old dug-outs. Without any doubt, these canoes were used to take their passengers to connect with the Winslow, or Roaring Creek, trail. Travellers would then go down to Harrison Lake and from here to Port Douglas [on Little Harrison Lake], then by trail and portage to Lillooet and further on into the Fraser River country.

The Kwantlen people Charlie Miller knew, even early in the 20th century, were a vestige of a larger nation. The home of their ancestors in the Stave had been changed by the arrival of Europeans.

Even before Europeans—known to the First Nations people as Xwelitem—set foot in the area, "passive" contact had already taken place. European tools and clothing had made their way into western North America along trade routes that extended to Mexico. Then, in 1782, a smallpox epidemic entered Stó:lō territory along those same routes. The results were devastating. The Stó:lō elder Simon Pierre of the Katzie First Nation, near Maple Ridge, gave this account of the epidemic:

A photograph of a group of Stó:lō taken by Frederick Dally, about 1860.

The news reached them from the east that a great sickness was travelling over the land, a sickness that no medicine could cure, and no person could escape. Terrified, they held council with one another and decided to send their wives, with half the children, to their parents' home, so that every adult might die in the place where he or she was raised. Then the wind carried the smallpox sickness among them. Some crawled away into the woods to die; many died in their homes. Altogether about three-quarters of the Indians perished . . . If you dig today on the site of any of the old villages you will uncover countless bones, the remains of the Indians who perished during this epidemic of smallpox.

Tony Dandurand, a member of the Kwantlen nation, explains the impact of smallpox on the Kwantlen:

Habitation in the lower Stave was more permanent, which is why we were there at the time of contact and why

THE KWANTLEN WAY OF LIFE

The Cheers, Fidels, and Gabriels were the principal Kwantlen families at the Whonnock and Stave River Reserves in the early 20th century. This photograph shows Mrs. Cheer and her grand-daughter at Whonnock around 1914.

FOR THE KWANTLEN PEOPLE of the Stave River, life revolved around salmon. The majestic fish was a food source, provided the focus of much daily work, and featured prominently in Kwantlen art and legends. The Stave River was a tremendous fishing ground, both at its confluence with the Fraser River and at natural fishing locations such as the impassable falls. Although everyone had a right to salmon, extended families owned traditional fishing locations. The salmon were netted, trapped, and speared. They were smoke- and air-dried on large communal racks or in smoke houses, and they were stored in cedar baskets for later consumption.

But salmon were also a commodity: the Kwantlen were not only fishers, they were traders. Stave Lake was part of a network of trade routes that linked the Aboriginal peoples of the coast, Washington state, and the British Columbia interior. The pre-European "provincial economy" was commodity-based and focused on resource redistribution rather than money. The five species of Pacific salmon were only one product line; sturgeon (which were harpooned or netted) may have been just as important. The Kwantlen were also regarded as "tireless hunters," and their animal skins were an important trade item, as were dried shellfish, berries and nuts, goat and dog hair blankets, bulrush and cedar mats and baskets, eulachon oil and dried eulachon, tools, and dug-out canoes. The various First Nations also raided and warred with each other. Slavery was part of the economy and culture and offered protection against inbreeding. The Kwantlen families were Halkomelem speakers and part of the larger Stó:lō (or river) people. They lived in permanent, horizontally planked, shed-roofed longhouses that housed both individual families and groups of families. They also maintained "winter" camps farther up the lake that were used for spiritual purposes, including ceremonies that could last for months.

Smallpox struck the Stó:lō in 1782 (before direct contact with Europeans) and again in 1824, 1848, and 1862; influenza and alcoholism became huge and deadly problems, as they were for Native peoples throughout the Americas. (Tuberculosis and venereal diseases were already present in North America at European arrival.) The terrible results of the epidemics directly affected the Kwantlen people and probably hurried the process of designating reserves in the area by 1879. The Kwantlen were affected by these diseases to such an extent that they number fewer than two hundred people today. The people of the Kwantlen First Nation are dedicated to preserving their legacy, pursuing land claims, and rebuilding their culture.

we were allotted reserves there . . . There was a village at the mouth called Sky'ux, which means "the place of all dead," because of the smallpox epidemic that came up through the trade routes in the 1780s. Kwantlen people used to number in the thousands and were one of the main tribes of the lower Fraser. There are only 140 of us now.

Not many years later Europeans appeared on the Fraser, and their coming ushered in a new era.

THE EUROPEAN ARRIVAL

The European presence in the lower Fraser River valley is usually dated to 1808, when Simon Fraser and his powerful North West Company paddlers descended the river from the rough outposts of New Caledonia to the coast. In his journals, Fraser does not mention the river that would become known as the Stave, but he was riding powerful spring run-off currents and feared for his life. The goal of his journey was to determine whether the river that would come to bear his name was the Columbia. It was not; in fact, the huge river, with its enormous whirlpools and punishing rapids, was not considered to be of much use at all. It would be sixteen years before Europeans appeared in the Stave area again.

In 1821, the North West Company merged with its arch-rival, the Hudson's Bay Company (HBC), to form a much larger entity that retained the latter name. In 1824, HBC chief factor John McLoughlin sent an expedition under the leadership of James McMillan from Fort Vancouver (now Vancouver, Washington) to reconnoitre a fur trading post on the lower Fraser. A clerk, John Work, was part of this expedition, and it was after him that the Stave River was named Work's River. Fort Langley was established by the HBC about three kilometres downriver from its present site in 1827. The Europeans had arrived, this time to stay.

Unfamiliar with the territory, the new arrivals were in many ways dependent on the First Nations people. This was certainly true for Simon Fraser, who had cached his supplies above Hell's Gate and relied on the Native people for food and directions. Later, the Kwantlen and other local First Nations were not always inclined to do the bidding of their new neighbours. The HBC then brought in Kanakas from Hawaii, workers from Chile, Métis from the Prairies, and other groups to work as labourers.

Archibald McDonald, the chief trader at Fort Langley from 1828 to 1833, quickly determined that the teeming salmon resource of the Fraser had tremendous trade value, especially because the Kwantlen people were already expert at fishing and processing the product. In a letter to HBC governor George Simpson, he detailed a significant problem and proposed a solution:

In my last Communication I touched at some length on the prospect of curing Salmon at this place, as an additional source of Returns, and I have the satisfaction to inform you that the experiment of last Season completely proved the theory; the fish, it is true, arrived late—indeed after we had almost given up hopes of making anything at all of it; but from the 20th of August to the 13th of the next month we were fortunate enough to procure upwards of 15,000; enough to make up more than 200 Barrels, which in that very short space we contrived to do, into nearly that number of casks of our own making, with means so imperfect, however that I fear from the sample that remained with ourselves, the first Cargo will not stand the Test of a foreign market, and trust by the next Season, we shall be provided with a good Cooper, that will know something of fish curing.

left:
John Work was born in 1792 in Ireland. He joined the Hudson's Bay Company in 1814 and first came west in its service in 1822. Work served the company in Fort Vancouver, Fort Simpson, and Fort Victoria, becoming chief factor in 1846.

right:
Archibald McDonald joined the Hudson's Bay Company in 1820 and arrived in Fort Langley as chief trader in 1828. McDonald's journals form an important record of fur-trade life.

Looking north and east up the Fraser River from Fort Langley in 1862.

As McDonald wrote, the problem with shipping salmon was making barrels suitable for transport overseas. The first efforts were failures; the barrels leaked and the salmon spoiled before it arrived at its destination in Hawaii.

In 1830, an excellent source of white pine—a wood suitable for making staves for barrels—was discovered just upstream and across the Fraser, where another river joined the larger flow. This was Work's River, which would be renamed the Stave River because it was the source of the barrel staves.

Beyond survival and the fear that one or more Native groups might wipe out the fort at any moment, barrel-making (and the problems relating to it) was a consuming problem, as Archibald McDonald's diaries from 1829 and 1830 show:

MONDAY [SEPTEMBER] 14TH. *It being very desirable to prepare a quantity of Staves here in Case we may undertake a more extensive fishery another year—4 men were sent back into the woods to examine the different qualities and returned with a fair Sample of the white pine—Commonly Called in this Country* le Pin Blanc [Pinus monticola—*western white pine*] *and not at all Common in this part of the world in the absence of Oak however it may answer our purpose . . .*

SATURDAY [SEPTEMBER] 26TH. *With four men & and a Small Canoe I visited our people in the woods up the river—we have now about 1000 excellent Staves of the kind—the wood looks uncommonly fine & well adapted for all household furniture—'tis a pity we are not in the way of Securing a quantity of it for exportation—The Country is too level for a Mill Seat & the nature of the bottom I believe not favourable for dams neither is the wood itself very Common . . .*

MONDAY [FEBRUARY] 15TH. *Four men Still after the Staves—and four more sent to Cut hoops—this is an article of which we use an incredible quantity from want of the requisite means to go on with—An iron Hoop at each end of the Cask would greatly facilitate our Cooper work, but of every thing in this way we are perfectly destitute.*

Indeed the man now at that work is himself far from being an efficient hand—Since the 18th of last month assisted by another dressing his Hoops—they have only turned off 20 very indifferently made Casks of 25 Galls.—A Cooper we must have, if it is at all intended to go on with the Salmon business.

MONDAY [APRIL] 5TH. *Sent 6 men off to the pines, about 3 miles up the [Stave] river, where the one half will employ themselves rising Staves & the other half Squaring a few fine logs which all we Shall easily get home when the water is high in the Summer.*

FRIDAY [APRIL] 16TH. . . . *In the afternoon I visited our men in the woods—They have now upwards of 3500 Staves with a proportional Stock of Heads.*

McDonald's men split bolts from logs on the banks of the Stave River and floated them across the Fraser, where they were milled into staves and transformed into barrels by the trained coopers who were finally brought in. The barrels were also used to ship cranberries and hazelnuts. At first, the trees were logged at the confluence of the Stave and Fraser Rivers, but in time the HBC men moved further upstream. Logging, then, was the first European industry along the Stave River.

EUROPEAN SETTLEMENT

By the late 1850s the HBC's presence at Fort Langley had resulted in the scattered settlement of retired employees and their mostly Stó:lō wives on both sides of the Fraser River. In the Stave River area resided permanent settlers such as Robert Robertson, who took up land next to the Stó:lō settlement from which Whonnock got its name, and William Cromarty, the cooper at Fort Langley, who built a home at Glen Valley, on the south side of the river. Beginning with the Fraser River gold rush in 1858 and the subsequent Cariboo gold rush, these scattered settlers were gradually augmented by disenchanted miners and other mostly short-term settlers. A few of these settlers ended up in the general vicinity of the Stave River; the voters' list for British Columbia in 1874 mentions "Ephraim Pollock, occupation logger" and "Marcus Cox, occupation farmer" residing in the area.

The 1860s and 1870s saw only scattered settlement in the Stave River area, but various government initiatives would eventually have an impact there. One of these initiatives—undertaken first by the colonial government and later by the provincial government at the behest of the federal government—was the establishment of Indian reserves during the 1860s and 1870s. In 1864, an eight-hundred-hectare parcel of land was set aside at Whonnock, just west of the mouth of the Stave River. This area was later reduced to a mere thirty-seven hectares; the final allocation of reserves did not happen until 1879, when Indian Reserve Commissioner Gilbert Sproat allotted the Kwantlen a total of 560 hectares spread over eight reserves. Three of these were along the Stave River where, Sproat noted in his diary, the Kwantlen "much wished for their land, and they said if agricultural land could not be got for them,

Sir John A. and Lady Macdonald (right), taking in the scene on July 24, 1886, at the CPR's bridge over the Stave River near its confluence with the Fraser River.

they would like timber land, as they cut logs for mills, and also firewood for steamers."

Following the completion of the Canadian Pacific Railway (CPR) in British Columbia, settlers began to pour into the province. In the Stave River area, rail service began along the mainline in 1883–84. The east-to-west line was completed in 1885, and passenger service began in 1886. A station for the area was established at Whonnock. Now people began to take up land around Whonnock, across the Fraser, and along the Stave River. Not too much later, settlers moved further north, above the falls, and west of the river. Much less settlement took place east of the river; there were scattered holdings, but most of the land was reserved by timber interests.

All of the land in the settlement area fell within the "railway belt"—a strip of land, sixty-four kilometres wide, land granted by the province of British Columbia to the federal government to help pay for the construction of the CPR. Within the belt, land—up to sixty-four hectares (160 acres)—could be purchased from the Crown for a dollar an acre (later increased to five dollars) by settlers who met criteria set down by the federal government. Locally, this system was administered by the Dominion Lands Office in New Westminster.

For these early settlers, the centre was Whonnock. In its first twelve years, the school in Whonnock was known as Stave River School. In Whonnock were the store, the post office, and the railway station for the entire area. Once a person got to Whonnock, transporta-

Hard at work in the woods near Stave Falls, around 1912.

tion was abundant. Trains ran several times a day into Vancouver and New Westminster, which, along with other communities up and down the Fraser River, were served by paddle wheelers that could pick up and deliver freight and passengers almost anywhere along the river. At Port Hammond, passengers connected to a steamer service to and from Victoria.

The Stave River was the highway for settlers. For the first three kilometres above the Fraser, there was ready access to land, but then the Stave narrowed and flowed through a gorge. For those who settled above the gorge, access was more difficult, and the construction of trails and roads quickly became a lively topic in local politics.

Stan Carter, a resident of the area, described how his family made its way from Whonnock, where they got off the train, to his uncle's homestead at the corner of Dewdney Trunk Road and Bell Road, just west of Stave Falls, in 1890:

There was no wagon road or other means of transportation for our long track from Whonnock to Stave Falls. So my father placed my brother and I in a "gunny sack," cutting holes in the bottom and having placed our legs through these holes placed us on his back while Mother making the best of it followed along behind bearing such necessities as she could handle. It meant hours of following blazed trails and crawling over wind fallen trees some to the diameter of eight to ten feet on which niches were cut to get a toe hold.

Access to and through the area improved somewhat in the 1890s, when settlers built a track from Whonnock Creek to the mouth of the Stave. A wagon road was started in 1897 that followed the western river bank from near the mouth of the Stave River to Morgan's Slough, just north of the falls. To cross the river and get to Mission, travellers had to go all the way up to Stave Falls and cross on a log bridge. (The Red Bridge, at the site of the Ruskin dam, near Ruskin, was completed in 1909, saving travellers the trek to Stave Falls.) The Dewdney Trunk Road from the Blind Slough to Mission was finished in 1911, but the connection with the Stave River road was poor.

Whonnock's general store was a focus for the settlers in the Stave area. L.C. York, standing on the front porch of the store with his daughter Gwendolyn, was store owner and postmaster from 1894 to 1906.

Most people who settled along the Stave were not, strictly speaking, farmers. They often had a large garden and planted fruit trees, but they counted on other activities, principally logging, to keep their families fed and clothed. Hunting and fishing were also important. When Stan Carter's family took up a homestead north of Stave Falls in 1892, they found that deer, grouse, and fish were plentiful:

You may find deer within twenty minutes of hunting. Blue grouse used to come close to [our] dwellings and feed on clover like geese. We used to shoot them with a bow and arrow. As for fish—it was no problem, you could get more than you could pack home in an hour. There were Dolly Varden, Cutthroat, lake and brook trout. We [took] Dollys up to almost three feet. Catching small trout was child's play; we would take a piece of string, tie a knot and bend it [a common pin] like a fish hook and with a piece of red wool wrapped around it for bait you could fish all day without the bother of changing fish lures.

Settlement also took place up toward the southeast arm of Stave Lake, along Cascade Creek, northeast of Mission. J.A. Kirk, employed by the federal government in 1890 to survey the valley, described it as "sheltered by the mountains to the north, east and west, and well exposed to the sun." Kirk thought the valley would "probably prove a good locality for fruit culture." A journalist enthused in 1891 that this valley had "passed from its forest stage to the home of intelligent workers."

No more will be heard the growl of the bear or the stealthy tread of the panther . . . to live in a valley without high wind or thunder storms, in sight of these great mountains ranges which shut in this land of marvellous beauty will soon be deemed a blessing indeed. The development of the country has just begun.

Joseph Robson Hairsine settled on the east side of Stave River in 1881. His wife, Mary, was a daughter of William Cromarty, cooper at Fort Langley. She died in 1896 and lies buried near the Ruskin dam. This photograph was taken around 1900.

THE STAVE'S FIRST INDUSTRY

In 1882, the *Daily Columbian* commented on one of the most valuable resources of the area:

Stave River station marks the best timber district on the river besides constituting a natural centre for an important agricultural district. Upon the slope extending from the bank of the Fraser to the mountains (ten to fifteen miles) stands as fine a lumbering forest as can be found anywhere.

Just as the forestry resource had been vital for the early HBC traders, lumber became the primary industry of the Stave in the late 19th century. At first logging was almost entirely done with manual labour. Only the best trees were taken—giant firs and cedars that required springboards to be cut into them in order to reach a manageable height for sawing. When the industry began, the trees were felled into the lake or river, or oxen and horses were used to drag the logs to the water.

The lumber industry took various forms. There were independent hand-loggers, family-run logging operations, and sawmills, large and small. Ruskin owes its existence to the sawmill established there by the Canadian Co-operative Society in 1896. When the co-operative lost control of the mill in 1899, it was taken over by E.H. Heaps, who operated it until about 1913. One of the family-run logging shows was that of the Carter brothers—Foster, Edward, and Norton—who started logging in the 1890s, cutting timber in various places along the river, above and below the falls.

One of the early hand-loggers on the Stave was Jack McConnell. His niece, Annie Hill, described her uncle's annual round, which took him from his small farm, where he planted potatoes in the spring, up the Stave River by dug-out canoe to the lake, where he cut shingle bolts, then back down the river with his bolts in early fall, and on to New Westminster, where the bolts were sold at the mill and he collected his money, bought new clothes, and returned to visit with his niece's family at Silverdale, before returning to his own home for the winter. McConnell was renowned in the neighbourhood for the many dogs he kept; he was reputed to be able to call each one of his nearly fifty dogs by name.

Logging is still a tremendously important part of the local economy. But even as the lumber industry was taking shape in the 1890s, other entrepreneurs were beginning to look to the river itself and to the power of the moving water as a resource to be exploited.

RUSKIN: A UTOPIAN EXPERIMENT

WHEN THE ENGLISH CRITIC and social theorist John Ruskin began to expound his theories in the 1860s and 1870s, the banks of the Stave River in faraway British Columbia would never have crossed his mind. Ruskin, a man born to wealth and privilege, wrote about the need for old age pensions, national education programs, and organized labour. He preached the disavowal of machinery and encouraged communal living and profit sharing.

Inspired by Ruskin's ideas, the founders of the Canadian Co-operative Society procured land on the west bank of the Stave River in 1896. There, they proposed to run a logging operation and sawmill collectively, sharing all profits and creating a perfect state of socialism. The society had 54 members in 1897, 35 of whom worked in the mill and its associated lumbering operation.

John Ruskin might not have approved of the sawmill, but the Canadian Co-operative Society had to make compromises to survive. The society purchased equipment from E.H. Heaps & Company, and Ruskin Mills was founded. Local settler Charles Whetham was an early supporter (but probably not a member), and he began a club to discuss Ruskin's social principles. The community's families lived independently, but they did share a communal garden. The Ruskinians built a small store and a school for the children, who were taught by Cora Tingley, whose salary was paid by the province.

Unfortunately the bold experiment foundered on the hard shoals of business reality. The society had trouble getting access to timber, and the sweltering summer of 1898 saw the flow of the Stave River reduced to a trickle. As a result no one could boom logs down to the confluence. The co-operative ran out of money and credit, and E.H. Heaps & Company foreclosed on the mill equipment. Today, all that remains of the original co-operative settlement is the enduring name, Ruskin.

Edward H. Heaps's new mill at Ruskin in 1912, after fire destroyed the first one. Another operation starting at that time just up the river on the west side was Stoltze's, remembered now as a shingle mill.

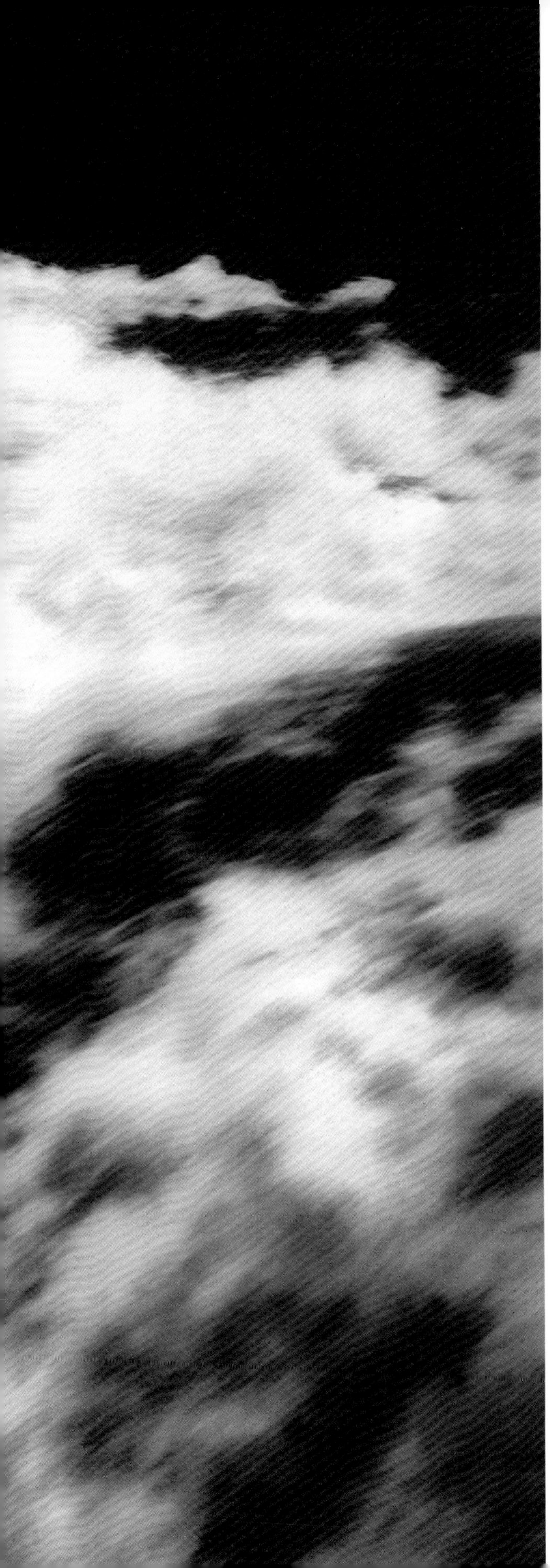

The Stave River shows its force below the falls in 1910. The total fall of the river from the mouth of the lake, at low water, to the Fraser River was about sixty metres.

CHAPTER TWO

Visions of Power: The Corporate Dream

TO AN ENGINEER or businessman at the beginning of the 20th century, Stave Falls looked as if it were created just to generate power. As a local newspaper noted in 1910, "The formation of the river at the point where the plant is being erected lends itself admirably to the work in hand and makes one wonder if an all wise Providence did not have an electrical power plant in mind when fashioning this part of the face of the earth."

All the elements were in place. There was Stave Lake, a natural reservoir fifteen kilometres long and one and a half kilometres wide, fed by a huge watershed of rivers and glaciers. From the lake, the river flowed to Stave Falls, where granite canyon walls narrowed sharply. Deep below the riverbed was rough bedrock, perfect for preventing the sliding of any massive structure that might be built on it. In the "V" of the canyon, the granite walls narrowed even further in the downstream direction, so it seemed that a dam wedged in the gorge would only jam more firmly into place over time.

The drop of the falls created a natural "head," over which water fell at a tremendous velocity. If one were to raise the level of the lake and channel the water through a flume (or "penstock"), that head would increase. With the construction of a powerhouse, electricity could be generated by the energy of the falling water.

To the west, in Vancouver, and to the south, in Washington state, were burgeoning cities and industries that would be hungry for power. Between them lay some of the best agricultural land on the continent. The Fraser Valley was home to many newly created communities, all of which would soon need a reliable source of electricity. In the years leading up to 1909–10, when the Stave Falls project was started in earnest—after a few false starts—this potential seemed obvious.

Remarkably, only a generation earlier, no one knew just how large amounts of electricity could be transmitted even more than a few kilometres to meet the demands of urban centres. The vision of power from the Stave was based on the obvious power potential of the falls and the river, and it was simply assumed that the technology would catch up to the potential. But it would take years of engineering gymnastics, corporate intrigue, cash shortfalls, and political wrangling before the vision was first transformed into reality—and longer still before the hydroelectric potential of the Stave River watershed was fully exploited.

THE HOLY GRAIL OF TRANSMISSION

The growth and development of electricity in late 19th-century North America was explosive. In 1870 no electricity was being generated anywhere on the continent, but by 1880 there were 5,453 production units operating, including 343 in Canada. At first most electricity was generated by steam. Water power was used mainly for mechanical purposes, such as driving a water wheel that was connected to a drive shaft that in turn directly powered machinery. Only later was it recognized that the force of moving water could power turbines that turned generators, resulting in hydroelectricity.

All early electricity-generating operations in British Columbia were small and local. Victoria had a limited electric lighting system driven by a twenty-five-horsepower steam engine in 1883. On the Lower Mainland, the Vancouver Electric Illuminating Company operated a steam plant at Pender and Abbott Streets starting in 1887. After the appearance of Thomas Edison's affordable and practical incandescent light that year, electric lighting was installed in more homes and businesses. Electric streetcars started operating in both cities in 1890, and in 1891 the first interurban rail service in British Columbia—between Vancouver and New Westminster—was inaugurated.

By the early 1890s the technology to generate hydroelectric power on a large scale had been developed, but transmission and distribution remained problematic. No affordable method existed to transmit large amounts of electricity from distant power sources to the large cities where it was needed. At the time, an epic battle between two of the luminaries in the field—Thomas Edison and George Westinghouse—was being waged in the United States over the best technology to do the job: Edison's direct current (DC) or Westinghouse's alternating current (AC). DC technology may have been more limited, but Edison conducted a high-profile and bitter campaign to promote it as the standard. Such was the great inventor's reputation that he almost won.

Westinghouse believed in AC for power transmission. He had purchased the patents and services of Nikola Tesla, the eccentric 19th-century Serbo-Croatian genius who had invented a revolutionary "polyphase" system. Tesla used the principle of a rotating magnetic field produced by two asynchronous alternating currents that made it practical to send high voltages for great distances along small wires and at reasonable rates. The higher the voltage, the lower were the losses of power. As it turned out, Westinghouse was Edison's match when it came to promotion—and his technology was superior.

The proving ground for the large-scale generation and transmission of hydroelectricity in North America was Niagara Falls, a place that epitomized power. Electricity could be generated at the massive cataract, but how would it get to "distant" Buffalo, New York, the closest major city? In 1889 the prospectus of a predecessor to the Niagara Falls Power Company announced:

It is conceded by leading practical electricians that it would be entirely practicable now to light the city of Buffalo (distance 20 miles) with power furnished by Niagara Falls, and the opinion is rife among scientific men that ways will be found in the near future for trans-

Vancouver's Hastings Street, aglow with electric light in 1914. Brightly lit streets were a hallmark of civic progress in the early 20th century.

mitting this power to much greater distances, and for using it in many new ways.

In fact it took six years, an international competition, and many proposed technologies before this prediction came true. Only in 1895 was Niagara power—at eleven thousand volts—finally transmitted to Buffalo, using an AC system.

In southeastern British Columbia, a company was established in 1897 to build the first major industrial hydroelectric installation in the province. Power generated at Bonnington Falls, on the Kootenay River west of Nelson, supplied a mine fifty kilometres away at Rossland with a transmission line operating at twenty thousand volts. When the job was completed in late 1898, this installation was the longest and highest-voltage transmission line in North America. The record was short-lived; new facilities steadily came on-line across the continent, providing higher voltages over longer and longer distances. Early in the 1900s the Mexican Light and Power Company built a 450-kilometre line at 100,000 volts, and by the 1930s the Hoover Dam on the Colorado River supplied a line of 480 kilometres at 287,000 volts.

facing page:
Maps made by the municipality of Mission and Western Canada Power were used to facilitate negotiations with land owners, to explain the project to potential investors, and, of course, to plan the technical details of the project.

EARLY VISIONS OF STAVE POWER

The power potential of the Stave River was recognized during the 1890s by a number of companies and individuals, but it is not clear what rights they held or how far their efforts went, beyond the planning stage. The earliest claim to the power of Stave appears to have been laid by the British Columbia Timber Lands and Electric Company, whose 1893 certificate of incorporation details an intent to "purchase, lease or otherwise acquire and hold lands, timber leases, booming and rafting privileges, and water privileges" in British Columbia. Its focus was strictly Stave Falls. Little is known about the company and its officers—John Wilson, the "mill manager," and John William McCrae and Dennis Murphy, "shipping agents" from Ottawa—or precisely what these three men planned to do at Stave Falls. In 1899 the Crown timber agent for New Westminster, James Leamy, recalled that the company had "been maturing some scheme for utilizing these falls," and that it had acquired a half-section of land; but he could find no record of the company having obtained water rights.

The Stave wasn't the only river to catch the eye of the aspiring power moguls of the 1890s. There were certainly other local possibilities for generating hydroelectric power. In the early 1890s the Vancouver Electric Illuminating Company planned to harness the Seymour River and use run-of-the-river technology (in which no reservoir is created) to power a new streetcar system—but the task of transmitting the energy fifteen kilometres from the North Shore seemed monumental. Later, the British Columbia Electric Railway Company of Canada (BC Electric) would secure water rights to a variety of prospective power sources, including the Capilano, Indian, Upper Pitt, Chilliwack, and Coquihalla Rivers. But none of these was ever developed; no water sources were better suited to generating power than Stave Falls.

IF AT FIRST YOU DON'T SUCCEED . . .

British Columbia Timber Lands and Electric Company was not alone in its interest in the Stave. In 1895 a group of Vancouver men incorporated a new company, the Stave River Electric and Power Company. The original directors, "James Beatty, of the City of Vancouver, prospector, James Augustus Wood, of the same place, capitalist, and Charles David Rand, of the same place, real estate agent," were allowed by the incorporating statute to:

Take and use from the Stave River, in the District of New Westminster, in the Province of British Columbia, so much of the water thereof as may be necessary to obtain therefrom power for the purpose of generating electricity to be used either for electric lighting, motive power or other works of the said petitioners, or to be used for the operation of stationary machinery, or for any other purpose or purposes, or to be supplied by them to consume as a motive power for hauling, pumping, lighting, smelting, drilling, or for any other purposes for which it may be applied or acquired, with the power to the Company to construct and maintain buildings, erections, dams, ditches, flumes, raceway, or other works in connection therewith for improving and increasing the water privilege, and also to enter upon and expropriate land for a site for powerhouses, or for carrying the electric current underground or overhead and maintain tramways for the purpose of carrying passengers or freight from some convenient point on the Stave River to any point or points within a radius of fifty miles from the mouth of the said Stave River.

Given the times, the company's goals were ambitious. Power from the Niagara River—which would finally prove the viability of long-distance hydroelectric transmission—had barely started flowing to Buffalo. Yet, Stave River Electric had its sights set on supplying power to Vancouver, fifty kilometres away.

From the outset, the proposed new company was controversial. In January of that year Colin Sword, the member of the legislative assembly for Westminster-Dewdney, presented a petition to the legislature from the Mission City Board of Trade opposing the original private member's bill to appropriate Stave water. A petition from the Westminster & Vancouver Tramway Company also opposed the bill because it included a clause permitting the construction of a street railway. Perhaps as a result of these protests, nothing seems to

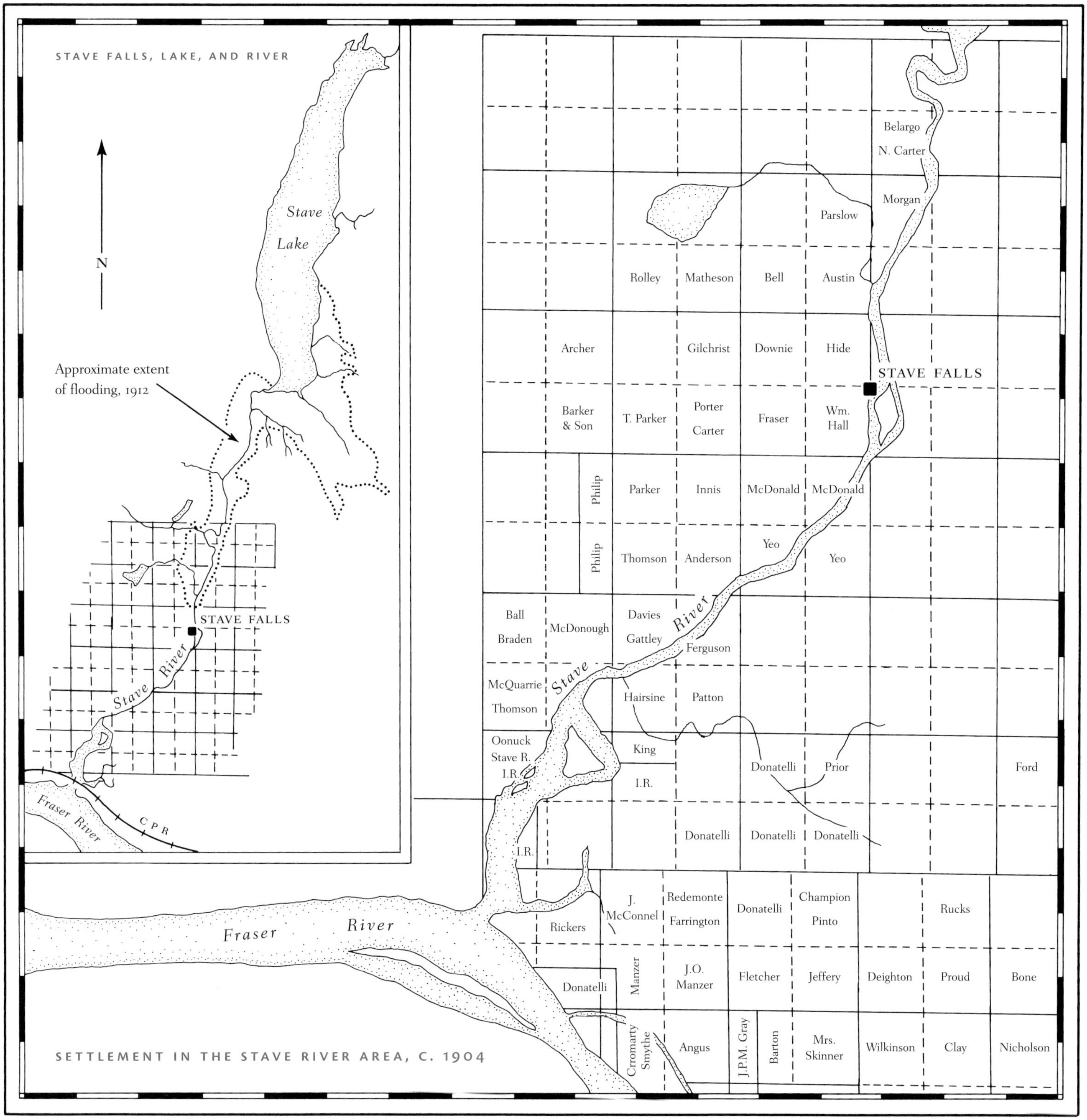
STAVE FALLS, LAKE, AND RIVER
N
Stave Lake
Approximate extent of flooding, 1912
STAVE FALLS
Stave River
Fraser River
CPR
SETTLEMENT IN THE STAVE RIVER AREA, C. 1904
Belargo
N. Carter
Morgan
Parslow
Rolley
Matheson
Bell
Austin
Archer
Gilchrist
Downie
Hide
STAVE FALLS
Barker & Son
T. Parker
Porter Carter
Fraser
Wm. Hall
Philip
Parker
Innis
McDonald
McDonald
Philip
Thomson
Anderson
Yeo
Yeo
Ball
Braden
McDonough
Davies
Gattley
Ferguson
McQuarrie
Thomson
Hairsine
Patton
Oonuck Stave R. I.R.
King
I.R.
Donatelli
Prior
Ford
I.R.
Donatelli
Donatelli
Donatelli
Fraser River
Rickers
J. McConnel
Redemonte
Farrington
Donatelli
Champion
Pinto
Rucks
Donatelli
Manzer
J.O. Manzer
Fletcher
Jeffery
Deighton
Proud
Bone
Crromarty
Smythe
Angus
J.P.M. Gray
Barton
Mrs. Skinner
Wilkinson
Clay
Nicholson

left:
Engineer Robert Francis Hayward came to the New World in 1894 from England. He held positions in Utah and Mexico before arriving in British Columbia in 1909. At Stave Falls he served as general manager, juggling the business and technical aspects of the project.

centre:
Electrical engineer William R. Bonnycastle designed the intake dam and powerhouse. Hired in 1906, he stayed with the project until 1911, when the design work was completed. He later worked on the Bridge River power project.

right:
Engineer Charles Anthony Stoess came to Canada in 1883. He had a long career in the West, working for the CPR and on an irrigation system of the Keremeos Land Company. Between 1899 and 1903 he carried out survey work and made development proposals for Stave Falls.

POWER VISIONARIES

FORMULATING A VISION of what was possible on the Stave River was only the first step. Someone had to take that vision and make it a reality. Along the way, many threads had to come together—on the river, in the legislatures, with the bankers, in engineering, and in society generally. How they came together reflected a vision not just for the river, but for 20th-century Canada.

The vision of generating electrical power from the Stave brought together a fascinating group of engineers, Vancouver businessmen, and big-time capitalists with interests in industries across North America and abroad. In the early years, before 1909, local entrepreneurs, including contractor and aspiring car manufacturer W.H. Armstrong, steel manufacturer and mine manager Sylvester Faulkner, and electrical machinery merchant and contractor George C. Hinton, provided leadership and funds. John Hendry's lieutenant, William McNeill, was also an important player: he was credited in the press with assembling the eastern Canadian capitalists who saved the project in 1909. Typically, the early engineers were retained to carry out specific tasks. Most, like Charles Anthony Stoess, the first engineer on the job, were civil engineers. One exception was William R. Bonnycastle, designer of the intake dam and powerhouse and a trained electrical engineer.

When Western Canada Power, a subsidiary of Royal Securities Corporation, took over the project in 1909, the web of interests widened. Royal Securities was run by Max Aitken (Lord Beaverbrook) and, later, Izaak Walton Killam, Canada's leading capitalists in the first half of the 20th century. It had extensive interests in the power industry across North America and owned Montreal Engineering, the firm that handled the Stave project. Robert Francis Hayward was brought in to manage the project fresh from Royal Securities' Mexican Light and Power project. With him, he brought resident engineer A.R. MacKenzie.

It is difficult to know what attracted these individuals to the project. Some wanted the challenge and excitement of working with a new technology. Others believed that the provision of electricity was a public good—people's lives would be better for it. For many, the idea that they were transforming otherwise wasted energy into something useful was very attractive. For yet others, the promise of profit urged them on. But for most of these men, the attraction was probably a combination of all these factors and more.

have come of Stave River Electric's plans. But at about this time the power of the falls began to turn the wheels of industry, albeit in a small way. According to a later newspaper article, in about 1896 a small group of settlers had "put all their available cash into a small sawmill, which derived its power from the falls."

The Stave River also attracted the attention of BC Electric, which, when it was formed in 1897, investigated the river's power potential. An account book for that year records an expenditure of $189.40 to explore the Stave and Sooke Rivers. The Stave report was prepared by James Garden, a civil engineer who would soon serve as mayor of Vancouver. Garden's report on the potential of Stave Falls proposed a relatively simple and low-cost facility that would use a pipeline or flume running from the falls to a powerhouse near Ruskin and would cost less than $200,000 to construct. Although BC Electric did not follow up on the report, the company would eventually become an important player in the Stave River drama.

Finally, in 1899, the firm that would jump-start the project was formed: the Stave Lake Power Company, whose three principals were John B. Ferguson, George C. Hinton, and W.H. Armstrong. The province approved the company's application to use the water of the river to generate electricity in April 1900. The certificate of approval permitted Stave Lake Power "to use water from the said Stave River . . . not exceeding seventy-five thousand inches, for the purpose of developing electrical power and supplying the same to any persons, companies or corporations, and at such points and for such distances as it may be found practicable to transmit electrical energy." The text concluded: "This certificate is issued upon the condition that all tolls or rates to be charged to consumers for power shall be fixed by the Lieutenant-Governor in Council, and may from time to time be altered."

Stave Lake Power's prospectus of 1900 was very optimistic:

The Company have secured by purchase all the rights to 75,000 inches of water at the Falls and the land adjacent. The value of these may be conjectured when it is stated that the Government charges alone have been nearly $10,000.00. Its value will be still further appreciated when it is known that there is no other stream on the Pacific Coast of Canada which contains so great a power, which can be developed at so low a cost. The position of the Falls on the River and the conditions of the country adjoining are such as to make the development the most economic possible.

There was a pressing need to supply the Lower Mainland with large amounts of power. BC Electric, which was run from London, England, was conservative and slow-moving. Although it was considering all sorts of potential projects for generating more power, turn-of-the-century Vancouver industries still had to rely on a patchwork quilt of private and public steam plants to supply electricity, and the city's largest industrial producers were desperate for more electrical power. The presidents of more than one hundred companies made it clear to Stave Lake Power that the sooner electricity

The letterhead of the Stave Lake Power Company shows the falls on the river, transmission lines, and a small electric motor.

could be transmitted from the falls, the better. They all made modest investments in the company. Stave Lake Power's prospectus declared:

The Company has received the endorsement of a large proportion of the leading business men in the two cities most interested [New Westminster and Vancouver], as will be seen by a reference to the list of subscribers of stock, and it has the cordial support of all classes of both cities.

One name among these many industrialists stood out: John Hendry, president of the Vancouver, Westminster & Yukon Railway Company and owner of numerous businesses, including the Hastings and Moodyville sawmills on Burrard Inlet. Hendry was so influential that when he moved to Vancouver from New Westminster early in the 20th century, it signalled the younger city's rise as British Columbia's business capital. From the outset, Hendry was a supporter of Stave Lake Power; later he would become its saviour.

Operations at Stave Falls began with a great deal of vigour, and there was much to do. The English- and German-trained engineer Charles Stoess was hired to survey the territory, create a record of river flows and volumes, locate a site for a hydroelectric facility, and propose the type of plant. He carried out his preliminary surveys in 1899 and 1900. At the same time, talks were started with British Columbia Timber Lands and Electric Company to purchase its land holdings and discussions with other landowners in the area began.

Negotiations for a franchise to sell power were begun with the City of Vancouver in 1901 and with the City of New Westminster in 1902. Although BC Electric already provided Vancouver with power for street lighting and had the concession for a street railway in the city, other municipalities and industrial contracts were fair game for Stave Lake Power.

Shortly after it was formed, however, Stave Lake Power ran into the problem that would bedevil it until long after the falls were dammed: money. The company simply did not have the funds to build a $2 million facility. To solve this problem, the directors hoped to find major partners to assume the financial burden, and the process of wooing potential backers began immediately.

A series of would-be suitors appeared. First off the mark, in 1900, was the Boston firm of Stone & Webster, which owned many generating facilities in the United States, including Puget Sound Traction, Power and Light Company, which operated most of the power plants in the Seattle area. Stone & Webster sent engineers to Stave Falls to report on the viability of the site for a major hydroelectric installation. The engineers began their site investigations on October 14, 1901. But on November 1, all operations at the falls suddenly shut down when Stave Lake Power realized that its prospective partners were charging all their expenses back to the Canadian company, which found itself having to pay a five-thousand-dollar bill. Also in 1901, C.H. McIntosh of the British American Mining Corporation—owned by a London-based parent company—proposed to purchase Stave Lake Power. But a series of stock market scandals and frauds put an end to this proposal.

Two years later, Stave Lake Power was asking $35,000 for its rights. Charles Baker of the Snoqualmie Falls Power Company in Washington state offered half that and proposed a partnership to form a distribution company, with himself as the major shareholder. But his father held the purse strings, and when the elder Baker passed away later that year, negotiations ended. In November 1903 a Vancouver businessman named Thomas Wilson offered to meet the asking price. Wilson had served as secretary-treasurer of the company and did some work at Stave Falls; but these negotiations ceased. Next, prominent Vancouver solicitor Osborne Plunkett stepped up to the plate. He may have been a front man for another group, perhaps even the City of Vancouver, but nothing came of these negotiations either.

Meanwhile engineer Charles Stoess persisted with his investigations, undertaking geological testing at the site. A half-century later, a local resident recalled the beginning of construction:

In the spring of 1903–1904 the Stave Lake Power Co. built a log house approximately where the Stave Falls public school is today [*above the dam site*]. *This was to house their surveyors and crews. A cable suspension bridge for foot traffic was then constructed at this point.* [*The*] *next building started about 1905.*

By 1905 Stave Lake Power was in a financial shambles and had neither generated nor sold a single kilowatt of power. John Ferguson resigned as treasurer amidst suggestions of financial irregularities. Into the breach stepped John Hendry and Sylvester Faulkner, another prominent shareholder. They paid the outstanding bills and lent the company fifty thousand dollars, secured by a mortgage on the land and the franchises negotiated to that point.

JOHN HENDRY, FINANCIAL SAVIOUR

Better than anyone, John Hendry understood the Stave's potential. He was most prominent among the original investors and had attended the first Stave Lake Power Company directors' meeting. Now, he was the chairman of the board and owner of 70 percent of the shares. Also joining the company as secretary was William McNeill, Hendry's right-hand man at the Vancouver, Westminster & Yukon Railway. In 1905 the company undertook work at the site in order to meet the terms of both the provincial and the federal water licences.

The new ownership heralded a new approach to the project's design. Charles Stoess had devised a plan that could be completed quickly and economically: a "run-of-the river" facility that involved a low weir across the Stave, funnelling water into a flume (or penstock), and a riverside powerhouse. This plan was similar to James Garden's proposal for BC Electric. It would generate far less power—four thousand horsepower in its first phase. It was a brilliant concept that could have been completed within a year or two for $500,000. Most importantly, it would have earned a profit in short order. Stoess, however, found the slow pace of work frustrating. He left Stave Falls to work on other large projects in British Columbia and Alberta.

Stoess was replaced as chief engineer in 1905 by William Kennedy Jr., who had a very different, and more ambitious, vision for the river. He recognized the brilliance of Stoess's plan but also understood that the industrial power needs of the Lower Mainland could only increase. Kennedy envisaged four turbines, each capable of producing four thousand horsepower, that would require a high dam and powerhouse. The price of Kennedy's plan was much higher, since this project would cost $2 million. The die was cast; this scheme contained the basic elements of the one that was eventually built.

Planning and replanning the project required time and cost money. John Hendry frankly assessed the situation in a letter to shareholders:

This year a start in actual construction had to be made in order to kept faith with the Government and Cities of Vancouver and New Westminster which required that a large amount, about $100,000, should be spent on the works and plant to hold the franchise and privileges which we have obtained from them. It was also necessary for the same reason to secure competent service in the preparation and carrying out of plans for the construction of the dam and power-house, and for this purpose the services of Mr. William Kennedy, C.E. of Montreal were engaged as consulting engineer and Mr. James Kennedy, C.E. as Resident Constructing Engineer. In order to raise the $100,000 necessary, the shareholders at meeting on August 23rd of this year authorized the execution of a mortgage for that amount to Messrs. Hendry and Faulkner.

John Hendry, Vancouver's leading lumber and railway baron, supplied the money needed to keep the project moving at various points.

Hendry's plea must have been successful, because the work begun in 1905 continued in earnest in 1906. The first things that had to be done were heavy repair

Charles H. Cahan's early enthusiasm for the Stave Falls project later faded as he struggled to raise enough money to keep construction moving.

work to the Stave Lake Road, the renewal of all bridges, the construction of a permanent camp, and the erection of two swing bridges.

In 1907, only months after work began on the sluice dam, a general panic in the financial world caused funds to dry up. The project was halted in its tracks, and two hundred men were let go. A local newspaper reported:

The action of the company in suspending operations at the lake has caused considerable wonder among the settlers in that vicinity, and some of the residents have openly expressed a fear that the partially constructed permanent dam and piers will be a total loss if a strong freshet occurs this season such as came down the Stave River in 1897.

Every working day that was lost meant less cash flow from power sales down the line. But although work at Stave Falls stopped, planning continued. Engineer William R. Bonnycastle was hired to design the intake dam and powerhouse. Late in 1908 construction started up again. Soon the sluice and wing dams took shape. Then, once again, the money ran out. And, once more, Hendry stepped in, lending the Stave Lake Power Company another $100,000.

John Hendry was no fool. His resources were so vast that he could have personally funded the entire Stave Falls project. By 1909 he had already put up $250,000—a huge amount of money at the time. Nevertheless, ever the canny businessman, he intended to maximize his profits from every aspect of Stave Falls. And no one could play the game better than Hendry. At the end of the year, he bought Sylvester Faulkner out.

At the same time, a new and dramatic development occurred. On July 1, 1909, a group of Montreal businessmen, led by Charles H. Cahan, purchased the Stave Lake Power Company and renamed it the Western Canada Power Company. The new company's owners were the Bank of Montreal, Royal Securities Corporation (the Canadian fundraising arm of Lord Beaverbrook's financial empire), and a small group of influential shareholders, including Sir Charles Hibbert Tupper, son of former prime minister Sir Charles Tupper.

Cahan was brimming with confidence, as his letter to Royal Securities in May 1909 shows:

After careful examination of the Stave Lake Power Enterprise, I am convinced that the development of the power at a moderate cost is entirely feasible, and that the market for power in Vancouver and vicinity can be developed on a profitable basis. I have found several legal difficulties, which it would be wise to overcome at once by the organization of a new Company. I have caused application to be made for the incorporation of a new Company with ample powers for carrying out this enterprise.

Many shareholders surrendered their Stave Lake Power shares for new Western Canada Power shares under a scheme that gave them one new share for two old ones. They had little choice, because the project was stalled. Hendry's $250,000 loan was repaid in full; he kept 17,300 shares and received two thousand more for his "good will."

Western Canada Power brought in engineer Robert F. Hayward as the new general manager. Hayward immediately became the public face of the Stave Falls project. He had previously been general manager of the Mexican Light and Power Company (in which Royal Securities also invested heavily) and arrived with a cadre of professional colleagues with whom he had worked in Mexico and the western United States over the previous ten years.

Hayward planned more changes. Now the Stave Falls power plant would house four 12,500-horsepower turbines, providing more than three times the generating power of Kennedy's scheme and twelve times the power of Stoess's. Although this required a still higher dam, Hayward and his colleagues believed the downstream profits would justify the investment.

Unreliable suppliers and the forces of nature delayed the completion of the project by almost a year, but the momentum never stopped. Western Canada Power signed new franchises for power with Maple Ridge, Pitt Meadows, and Coquitlam. Negotiations began with South Vancouver, Point Grey, and Richmond. Construction began on a receiving station at Ardley, in Burnaby, and on substations in Sapperton and downtown Vancouver.

In 1910 the Stave Falls Railway—a federally chartered line that claimed to be the shortest railway in the world—was built in four months. It ran just under ten kilometres, from the CPR line at Ruskin, on the Fraser River, to the construction site at Stave Falls. Even this pace was too slow for the accelerated progress of the work. Construction reports from the time indicate that work at the site was delayed probably because the line had not been put in sooner. In that year, Western Canada Power also built a provisional power plant in the middle channel, which powered three cableways and three derricks. Other equipment remained steam-powered, in case the generator should break down.

THE POWER MATING DANCE

BC Electric—the Stave Lake Power Company's largest and most important competitor—had also had the opportunity to purchase Stave Lake Power in 1908. The incentive was strong because Stone & Webster were still interested in the Stave and had the resources to become involved in a big way. Moreover, Stave Lake Power was a major obstacle to BC Electric's creating a monopoly in the Vancouver area.

BC Electric carefully monitored the progress at Stave Falls and took a keen interest in Stave Lake Power's, and its successor Western Canada Power's, efforts to raise funds. A BC Electric letterbook from 1908 and 1909 tracks the company's decision not to acquire Stave Lake Power and follows efforts by John Hendry to raise money for Stave Falls in London:

AUGUST 17, 1908 [*letter from London*]. *It is reported that Mr. John Hendry left for England to raise funds for water power.*

FEBRUARY 11, 1909 [*letter to BC Electric chairman R.M. Horne-Payne*]. *Our Co. has contradicted reports in papers that our Co. has purchased Stave Lake Co.*

MARCH 1, 1909 [*letter to R.M. Horne-Payne*]. *Mr. Sperling* [*BC Electric's general manager*] *does not think it advisable to be drawn into a newspaper controversy. Will make use of suggestion that we should put out to public that this opposition to our plans can only put us to needless expense and thereby raise cost.*

APRIL 1, 1909 [*from Mr. Sperling's diary*]. *Long conference with regard to various rumours that have been received about Mr. Cahan's proposed program for developing Stave Lake.*

JUNE 14, 1909 [*from Mr. Laursen's diary*]. *Issue of BC Gazette, June 10th contains License to Western Canada Power Co. Ltd. To carry on business as an electric light, heat and power company & to take over, among other things, franchises, etc. of other companies. This is the Co. which has been formed to take over the Stave Lake Co.*

JUNE 16, 1909 [*letter from London*]. *Directors have decided to break off all negotiations with Stave Lake people and assume attitude of utter indifference and disgust.*

The entries for these two years provide a glimpse into BC Electric's intense interest in Stave Falls and the level of intrigue often involved. The company's records are full of such shenanigans. But in fact, between 1899 and 1920, BC Electric and Western Canada Power (or its various incarnations) were drawn to each

Western Canada Power and its successors had a symbiotic relationship with the various logging operations in the valley, accommodating their operations on the lake and profiting from the traffic they generated on the railway. Abernethy and Lougheed, which had a large shingle mill at Stave Falls, was the largest logging operation in the valley until a forest fire wiped out much of their timber in 1929.

other by a similar vision of power from the Stave. Periodically, the two companies would come close to negotiating a partnership. At other times they simply conducted business with each other. And then, much of that time was spent in court.

Political pressure was also applied to fence in Western Canada Power. In 1910 BC Electric attempted to prevent Western Canada Power from building an interurban railway system from the Stave River along the north shore of the Fraser River, to Vancouver. The original federal charter for the Stave Falls Railway allowed the company to extend far beyond the valley of the Stave and to lay down lines throughout British Columbia. This was a huge threat to BC Electric's plans for an interurban system and also concerned a host of municipalities. In January the company contacted Vancouver-area municipalities to warn them about the threat to their control of their local transportation systems.

Within days, representations to Ottawa were made, local politicians were mobilized, and the bill authorizing the system's construction came under heavy fire, with the desired effect:

FEBRUARY 11, 1910 [*letter to BC Electric general manager Henry Sperling*]. *Bill passed Railway Committee of House but was so completely changed and amended as now to be quite harmless to us. This Co. will now be in a position to appeal to the Legislature of BC for a charter to build, but the power to grant or withhold will now rest with the Legislature of British Columbia.*

FEBRUARY 19, 1910 [*letter to London*]. *Western Can. Pwr. Co. Bill as now amended limits construction to line from works [at] Stave Lake to some point on line of* CPR *between Ruskin & Hammond.*

At one point, the BC Telephone Company tried to insist that Western Canada Power not be allowed to string its transmission lines on roads where telephone lines were already strung. It helpfully suggested a half-mile (0.8-kilometre) separation (even though the lines of BC Telephone and BC Electric, companies that had some corporate connections, were often strung alongside each other). As a result of the wrangling, Western Canada Power waited seven months in 1910 to get permission to construct its sixty-thousand-volt line from Stave Falls to Pitt Meadows.

As construction continued at Stave Falls through 1910 and 1911, the prospects for more contracts and sales continued to look promising, as this letter from Robert F. Hayward to Charles H. Cahan indicates:

At the instance of an electrical man named Rives and another man named Morrison, both residents of Bellingham in the State of Washington, I sent Mr. Nims and Mr. Montgomery to examine the district between Sumas and Bellingham and they report an extremely good looking field for power. Bellingham has 30,000 inhabitants and is indifferently supplied with light and power on a small scale by a water power plant owned by the Stone & Webster people.

Between Sumas and Mission there is quite a little business to be picked up, the principal of which is the Clayburn Brick Company. This Company will require 350 [horsepower] immediately with a development increasing up to 1,000 h.p. the near future.

Hayward concluded that it would be a good business decision for Western Canada Power to build a transmission line from Stave Falls over to Mission, across the Fraser River, and down through Abbotsford to Sumas, Washington, at the U.S. border.

Hayward advocated a deliberate policy of "selling power for industrial purposes and as far as possible staying out of the lighting business." The logic of this sales strategy was simple. It was cheaper and easier to distribute power to relatively few large customers than to build and maintain an extensive distribution system for many low-consumption residential customers.

POWER FINALLY FLOWS FROM THE STAVE

Power began at last to flow from one of the two generating units at the new facility in December 1911. The vision of power from the Stave River had finally become reality. By late January 1912, the second unit was online and the sixty-thousand-volt transmission line to New Westminster and Vancouver (with limited distribution facilities in the cities) was completed. That year a contract was indeed signed with the Stone & Webster subsidiary, Puget Sound Light and Power Company, to supply Bellingham with power for ten years. A transmission line was built to Huntingdon (Sumas), and Western Canada Power's first commercial contract in the Fraser Valley was signed with Clayburn Brick. Other contracts followed with the American Lumber Company, the Kootenay Jam Company, and a variety of lumber mills and sash and door companies.

Although the focus of Western Canada Power's marketing strategy was on industrial customers, it did not entirely ignore the domestic market. In 1911 it initiated an aggressive marketing campaign to win over the hearts and minds of potential domestic customers along its transmission network. Full-page advertisements were purchased in local newspapers. At first, many homes' only use for electricity was for lighting. It took some years before most houses were fully electrified, because few had adequate wiring, and many appliances were unreliable. But electric irons soon became popular because they could be plugged into a light socket, and Western Canada Power aggressively promoted Hotpoint irons during the Christmas seasons.

On the surface, the launch of a brand-new, state-of-the-art generating facility was cause for celebration. Yet, the completion of the project and the signing of contracts hid a dire financial situation. The years of financial turmoil had repeatedly left Western Canada Power on the verge of bankruptcy, and the stress created by completing the Stave Falls project had taken a toll on

Electricity was promoted as being cleaner than other fuels, easy to use, and good value for anywhere light was needed or motors were operated. This advertisement appeared in the *Fraser Valley Record* in 1912.

both the company and individuals. Even as power began to flow, Cahan found himself under terrific pressure:

I don't see any way in the world whereby I can raise any more money for some time, until we can show a revenue, and it will be necessary to have available $100,000 in cash on July 1st to pay the current interest on the $4,000,000 of bonds now issued.

I am placing these facts before you so that you may understand the financial difficulties in which we are placed, and we must somehow keep within this limit even if we have to resort to all sorts of temporary expedients in furnishing the cheapest possible connection to our consumers.

I have gone through the greatest possible anxiety during the past month and have been practically crippled in raising these amounts of money; so crippled in fact that I have despaired at times of pulling through.

Western Canada Power needed every bit of business it could get. But in 1913, just as there seemed to be light at the end of the tunnel, an economic downturn enveloped North America. Making the best of the situation, Western Canada Power in that year signed a contract with BC Electric's subsidiary, the Vancouver Power Company, to provide bulk power at one cent per kilowatt hour. Vancouver Power sold it back to BC Electric at one and a half cents, making a tidy profit. The terms of the contract were complex and deadline-filled, and the agreement would bring BC Electric and Western Canada Power to court many times in the coming years. World War I and Western Canada Power's inability to meet the contract's installation timetable for generating units 3 and 4 played into BC Electric's hands.

A long and detailed letter from BC Electric's chief engineer, G.R.G. Conway, to its general manager, George Kidd, in 1915 shed light on Western Canada Power's problems and, once again, raised the spectre of Stone & Webster. Conway reported that Western Canada Power general manager Robert F. Hayward had visited recently and admitted that the company was in serious difficulty:

He was reducing his staff to its lowest limits, and said that he might have to ask us for some assistance in the way of men temporarily during an emergency. He also stated that if things came to the worst with them, the Stone & Webster interests would be willing to take them over, but they would do nothing with the Stone & Webster corporation before putting the situation before our Directors.

Mr. Hayward said that owing to the financial conditions due to the war he did not think they could comply with the [BC Electric] contract so far as completing Power House No. 1 at Stave Lake to a capacity of 30,000 K.W. in 1915.

It seems to me inevitable that the whole question of the Western Canada Power Company will become not only a very serious one for themselves, but also for our Company, and I feel very strongly that everything should be done to prevent their Company getting into such strong hands as the Stone & Webster corporation. Recently the representatives of the Stone & Webster corporation have paid frequent visits to Vancouver, presumably on holidays, but I have no doubt they are considering the position of the Western Canada Power Co. at the present time.

I am not suggesting that the present time is an opportune one for taking any action, but serious consideration should be given to the situation created by the difficulties that the Western Canada Power Co. are now in, to prepare for the possibility of the present control changing hands.

As the decade wore on, both BC Electric and Western Canada Power reached the limits of their installed generating capacity. In a number of reports, BC Electric engineers made the company's options clear: develop new sources of power, purchase power from Western Canada Power, or buy the rival company and develop its Stave Falls plant. Ultimately, Conway's assumptions proved correct: Western Canada Power would eventually have to change hands. Although generating unit 3 was installed at the Stave Falls plant in 1916, the turbine for unit 4 was delayed because of the war and remained stored in a Swiss warehouse.

The company was reorganized in 1916, this time as the Western Power Company of Canada (WPCC). But reorganization did not solve its problems, which were succinctly itemized by J.I. Newell, BC Electric's electrical superintendent, in a memorandum on the power situation in the Lower Mainland: "Certain factors have in the past, and are still, preventing that company (WPCC) from getting proper returns on its investment." These factors included unprofitable contracts for the supply of energy in bulk, the absence of any new large power contracts in the past seven years, and an inadequate distribution system.

In Newell's opinion, however, WPCC's difficulties were not insurmountable. With the right management and improved economic conditions, it could pose a serious threat to BC Electric. Newell advised quick action, before the situation changed, to forestall a "protracted competitive struggle," the outcome of which might be detrimental to BC Electric.

Through 1919 and 1920, BC Electric and WPCC were involved in serious amalgamation negotiations. A final agreement was signed in December 1920, and the deal took effect on January 15, 1921. On February 12, a group of WPCC employees at the Carter-Cotton Building (now the Province Building) at Hastings and Cambie Streets in Vancouver boxed their files, packed their bags, and walked two blocks down Hastings Street to BC Electric's headquarters at Carrall Street. There, they set up shop on the second floor as a more-or-less independent unit.

Although WPCC became an independent operating unit within BC Electric, the corporate language at the time leaves no doubt that the union was a takeover—the end of the quarter-century-long battle to supply the Lower Mainland with power. (For corporate financial reasons that had to do with bonds, WPCC would remain a legal entity until December 31, 1949.) When the *BC Electric Employees' Magazine* reported on the new arrangement, it foreshadowed how the continuing vision of power from the Stave would unfold:

The British Columbia Electric Railway Company of Canada took steps recently to acquire control of the Western Power Company of Canada. The control of this company does not at the moment supply us with increased power, because both the Lake Buntzen plants and the Stave Lake plant are fully loaded, and as a matter of fact we have been taking [the] majority of the Stave Lake plant's output for a number of years.

Western Canada Power's Ardley receiving station in Burnaby used Stave Lake Power to illuminate its message.

The Western Power Company's plant at Stave Lake, however, is capable of further development by the installation of another 10,000 kilowatt generator and the construction of a second plant down the river. The additional generator must be installed almost immediately and will entail an expenditure of a million dollars, and the second plant will be necessary in a few years.

Before long, few vestiges of the Stave Lake Power–Western Canada Power years remained. Robert F. Hayward ceased working for WPCC in 1920 and died in 1924. In that year, only five office staff from Western Canada Power remained working at Carrall Street. Operating staff were kept on at the Stave Falls plant; in fact, a few names from the pre-1921 days remained on the seniority lists well into the 1950s and 1960s.

Today, the body of water below the Stave Falls power plant—appropriately named Hayward Lake—is the only echo of a time when BC Electric was seriously challenged as a power producer in British Columbia's Lower Mainland. The original vision of power from the Stave was a grand one that would require many years and much work to bring to completion.

THE ALOUETTE-STAVE-RUSKIN COMPLEX

WHEN JAMES GARDEN and Charles Stoess reported on the power generating potential of the Stave River, they didn't limit their vision to the falls. Both noticed an additional site on the river where power could be developed, further downriver, where it flowed through a narrow gorge before emptying into the Fraser River. This was the Ruskin site. The men who laid claim to the right to develop the Stave were careful to identify both sites in their applications for water rights to government. In fact, they prudently obtained rights from both the provincial and federal governments, since at the time it was not at all clear which level of government controlled the water rights within the railway belt.

The genius of this situation, as Western Canada Power's directors explained to their shareholders in the company's annual report of 1911, was that "by the construction of a dam of not more than 120 feet high, in a narrow rocky gorge, the same water can be utilized over again, and an equal amount of power can be developed by the installation of four 13,500 horsepower turbines. The company, is therefore, in a position to carry out hydroelectric developments on the Stave River, within thirty-five miles of the City of Vancouver, up to a capacity of 100,000 horsepower, as rapidly as the market demands the power."

A sluggish economy followed by wartime exigencies prevented Western Canada Power from building at Ruskin, although for a while, general manager Robert F. Hayward held out the hope that the company could manage to build a cheaper alternative. This involved constructing, one at a time, two smaller dams below the powerhouse, thus staging the development. When BC Electric bought out Western Canada Power in 1920, it also obtained the right to develop the site at Ruskin.

The icing on the cake for BC Electric was that it already held the rights to use the water in the nearby Alouette watershed. The company's original plans called for the construction of a flume from the Alouette Lake's outlet to a

facing page, left:
Beginning in 1928, the operators at Stave Falls also controlled the generator in the small powerhouse at the foot of the tunnel from Alouette Lake.

facing page, right:
When it was completed, the Ruskin powerhouse was described as "A New Giant at Your Service." Designed as a peaking plant, Ruskin was developed in several phases. A second 47,000-horsepower unit was added in 1938, and the third and final unit was installed in 1950.

above:
A view of the expanded Stave Falls powerhouse and tailrace after 1925. Notable changes included the construction of the outdoor switchyard, the westward extension of the powerhouse, and the installation of an elevator from the powerhouse to the intake dam.

site seven miles away on the Alouette River, where it would drop eighty-five metres to a 14,000-horsepower powerhouse. With BC Electric's purchase of Western Canada Power, these plans were replaced by a scheme that would see a dam built on Alouette Lake and water diverted through a thousand-metre tunnel from Alouette Lake to Stave Lake. Power would be generated at a small 12,500-horsepower powerhouse on the western shores of Stave Lake. In addition to using the Alouette water to run a generator, the extra water flowing into Stave Lake made it possible to add a fifth unit at Stave Falls.

Work began on the Alouette project in 1924 with the construction of a rail line to the site, and tunnelling commenced the same year. In 1925, workers began building the dam at the foot of Alouette Lake that transformed it into a reservoir. The powerhouse was completed and put into operation in 1928. Described at the time as a "marvel of finer efficiency," it was the largest unit in Canada "operating under automatic control." Although the plant was automatic, it was manned; for many years an operator lived at the powerhouse. Another employee took care of the dam on Alouette Lake.

In December 1928, BC Electric decided to proceed with the development of the Ruskin site. The plant was "ready for service" a remarkable twenty months later. While irreverent wags liked to refer to Ruskin as the "last supper," when its builders paused to reflect on their achievement in November 1930 they described it as one of conquest, "of bending the rushing torrent to the will of man." BC Electric was tremendously proud of Ruskin, not just for the speed with which it had been constructed, but also because it had been "built by Canadian workmen practically entirely of Canadian materials." It was a monument "to British Columbian and Canadian Enterprise."

A panoramic view of the installation of the fifth and final penstock in 1925. When it was first put into service, air trapped in the fifth turbine caused "severe vibrations from the penstock and the throwing off of man-hole covers . . . a column of water is sometimes thrown approximately thirty feet in the air." Corrective measures were taken.

CHAPTER THREE

Building the Dream: The Technical Achievement

BUILDING THE STAVE FALLS generating station—"a monument to human industry and enterprise," in the words of one observer—was a major technical achievement. The project employed some of the continent's top engineers and craftsmen and hundreds of dedicated labourers. It promised a tremendous boost to the industrial development of the Lower Mainland and helped to modernize much of the Fraser Valley, especially the less-developed north shore of the river. Although the first stage of the Stave Falls powerhouse was completed in late 1911 and power started flowing to customers early in 1912, work at the facility continued over the years. Many upgrades were made, right up to the end. The dams were raised, additional turbines were added, the generators' capacity was increased, and a technically challenging fifth generating unit was installed. Finally, in the year 2000, a new state-of-the-art generating plant replaced the original powerhouse.

The effort to wring more power from the Stave didn't stop at the falls. On the rugged far shore of the lake, a tunnel from nearby Alouette Lake was pushed through and a small powerhouse was completed in the late 1920s. Downriver, around 1930, a dam and powerhouse were constructed at Ruskin. Each new installation increased the power that flowed from the Stave River to the Lower Mainland, the Fraser Valley, and Washington state.

The years between 1895 and 1915 were a formative era for the technology of large-scale hydroelectric power generation and transmission. Being part of a formative era has its drawbacks: new technologies are often superseded by even newer ones. In one crucial respect, Western Canada Power was too early in the game. When the Stave Falls powerhouse was built, only horizontally mounted turbines could withstand the enormous pressure exerted by the falls. But soon after the two horizontal turbines were installed, a new kind of bearing was developed that allowed the manufacture of massive, low-maintenance vertical turbines. It would, however, have been prohibitively expensive to rebuild the powerhouses to accommodate vertical turbines. So units 3, 4, and 5, all of which were installed later, were relatively small and inefficient horizontal turbines.

Although the turbine technology may have been a little behind the times, the construction techniques at Stave Falls were very much up to date. When BC Electric built its Buntzen Lake dam and powerhouse (the other British Columbia megaproject of the day) between 1901 and 1904, the work was done almost entirely by hand, using techniques that would have been familiar to labourers centuries earlier. Picks, shovels, and wheelbarrows were mostly used for excavating. The powerhouse was built of stone and was held together with mortar mixed by hand in flat-bottomed skiffs. Boats were needed to bring construction materials to the site, which was located on the east side of Indian Arm, near the end of Burrard Inlet.

When Stave Falls was built less than a decade later, much had changed. Although the first work on the sluice dam's foundation was done by hand, power drills soon replaced sledgehammers and hand drills. When the tailrace (the portion of the river below the powerhouse) was excavated, a large steam shovel was brought to the jobsite by rail. Powerful steam locomotives shunted rail cars full of steel, gravel, and other construction materials. Overhead cableways moved con-

The Marion steam shovel and her crew, from a 1912 postcard. Grant Smith and Company won the contract to excavate the tailrace. Robert F. Hayward noted that "the work proved very difficult and the price at which the contract was taken was too low, but the contractors carried through as quickly as possible and finished it on July 10, 1911."

crete and large objects around the site. The powerhouse had a steel frame and walls of reinforced concrete, which was prepared by steam-powered, and later electric-powered, mixers.

Charlie Miller, who was a boy when Stave Falls was built, recalled much about the times in his memoir, *The Valley of the Stave*. He described the steam shovel:

The Marion steam shovel was a huge, heavy rig equipped with a power "crowder" which could force the bucket into almost solid material and remove a yard and a half at each lift. It employed a crew of sixteen, who on many occasions would work twelve hours daily. All overburden had to be removed first, then "coyote" holes were drilled into the hillside, loaded with black powder and dynamite, and fired. This brought down thousands of yards of gravel, and in short order the side dump cars were filled and taken by one of the . . . locomotives to the top of the hill overlooking the job site.

The construction of the railway line from Ruskin to Stave Falls was essential to the success of the project. The construction site was about fifty kilometres from Western Canada Power's corporate headquarters in Vancouver, a long distance in the early 20th century. The dam, powerhouse, and transmission lines were located in isolated and rugged terrain. Workers had to overcome foul weather, plagues of mosquitoes, and constant delays. Although the lumber for the formwork and the gravel for the concrete were produced in the area, almost everything else had to be brought in.

When Western Canada Power took over the project in 1909, the only way to move supplies to the construction site from the CPR station at Ruskin was along the old and treacherous Stave River wagon road, which was entirely unsuitable. A wagon made on average only one and a half trips a day because of traffic congestion on the road—other wagons were constantly bringing in machinery, food, and other supplies. Moving something like the twenty-tonne base casting for the turbine over the road and its wooden bridges was unimaginable. As Western Canada Power's general manager, Robert F. Hayward, noted: "Until the railway was finished, the work was greatly delayed, due to the difficulty of transportation. Without the railway the construction work could not have been carried to completion."

Looking north, up the tailrace, toward the site of the powerhouse in June 1911. The spoil train, used to haul away excavated rock, is visible in this photograph, which also shows the steel frame of the powerhouse.

Starting in late 1908, after a few years of starts and stops, three more years of steady work were required to get the first two units at Stave Falls up and running. Impressive as this pace was, the timetable was even shorter for the Ruskin dam and powerhouse, built two decades later, which were completed in two years. These projects profited from improved materials, construction methods, and financing—factors that also benefited Vancouver's Lions Gate Bridge, which was strung across Burrard Inlet in nineteen months.

At Stave Falls and later at Ruskin, crews of between 150 and 500 men lived on-site and worked for foremen who possessed single-minded determination. Work continued for six and sometimes seven days a week. Overhead wiring and arc lamps allowed round-the-clock work. Stave Falls was a high-profile project, and skilled workers and engineers wanted to be part of the job. Many had worked with Hayward on other projects in Utah, Colorado, and Mexico.

The construction crew poses for the camera on the downstream face of the intake dam, which was completed in September 1911. The cables and wires in the background were part of the system of derricks and hoists used to move materials.

Gravel for the concrete was dredged locally, and lumber for construction was logged and milled in the area. Both activities gave local industries a huge boost. Charlie Miller described one of the interesting side-benefits of the construction work:

Thousands of feet of form lumber were sawn at [the] mill at Stave Falls. When the cement forms were "stripped," this lumber was thrown into the river because of the hardened cement adhering to it and was carried downstream. It was eagerly taken out from the river's back eddies by residents who constructed houses, barns and other buildings from it, most of the embedded sand particles now having been washed off by the turbulence of the river's current on its downstream journey. I knew of a barn, which housed a team of horses, three cows, a hay mow, and storage for all implements, that had been constructed from this discarded lumber. This same man also constructed several cozy cabins on his property overlooking the lower river which he rented to sports fishermen and hunters in season.

The project's construction and design relied on tradition and the occasional cautionary disaster. In 1889 the great Johnstown, Pennsylvania, flood had killed more than two thousand people—the result of a burst dam. In 1911 another Pennsylvania dam, in Austin, failed and killed more than seventy people. Referring to this catastrophe and to the 1907 collapse of the Quebec Bridge, in which seventy-five workers had died, the New Westminster *Courier* warned: "Might it not be reasonable to suppose that engineers are liable to be influenced in their opinions by the wishes of a large corporation? Who knows? However, it is always wise to add a margin of safety to their estimates."

As a result, the federal government appointed consulting engineer John Freeman, who specialized in the construction of dams and waterworks, to monitor the project. A lively correspondence went on between Freeman and Hayward regarding various engineering issues. For example, a number of seams of soft material compromised the bedrock at the falls and Blind Slough. The dams were engineered to drain the seepage and control foundation uplift, and a number of large potholes were filled with concrete. The Stave Falls builders understood the physical principles well, even though much formal theory didn't yet exist. (Definitive research on foundation uplift was not published until 1961.)

KEEPING THE RESIDENTS HAPPY

One unusual aspect of the Stave Falls worksite was the excellent amenities provided for workers and their families. Because of the project's distance from Vancouver and the often-foul weather in the area, Western Canada Power built a substantial camp for them. This far-sighted approach seems to have produced good results. Robert J. Hayward commented:

The arrangement made for the comfort of the men and the price we are paying for labour is not only securing us the best labour in the country but has resulted in keeping the men steadily at work in all weather, and while we have had to spend more money than I had figured in fixing the camp for the winter I feel sure that we will more than get it back.

A year later, Hayward wrote:

In order to hold the men and prevent the continual coming and going, which at the early part of last year, prevented any good organization of the work, a considerable amount of money had to be spent in providing for the housing and comfort of the men. The result, however, has been that the Company's camp has a reputation of being the best camp in British Columbia and the Company has been able to obtain the pick of construction men.

A report from Hayward to Western Canada Power's directors as the plant neared completion gives an idea of the price of treating workers well:

The cost of labour has been very high; the lowest price paid for common labour being $2.75 per day of 10 hours, while mechanics and carpenters get from $4.50 to $5.00 per day. The right kind of labour could only be retained by providing food and physical comforts for the men that would not be thought of outside of British Columbia; and the cost of doing this has added from 10 to 15 per cent to the price of the labour.

The effort to retain good help was conducted in many ways. For example, the company did not take advantage of the large unemployed workforce during the winter months. Although it was common practice in those days to pay workers less in the off-season, Western Canada Power maintained the same wage scale year-round.

A small, temporary powerhouse containing a six-hundred horsepower generator harnessed the power of the river, making electricity available for both camp and construction use. The sluice dam, which would later be raised and filled to create the main dam, is visible in this picture, as are the falls themselves.

The company campsite was built, stocked, and maintained far beyond the standards of the day. A newspaper article described the camp as "quite a pretentious affair; the whole place is brightly lighted by electricity from the company's temporary power plant, and the 'Midway,' as the boys call the main avenue through the camp, in a measure resembles Columbia Street [in New Westminster] after dark."

The Vancouver *News-Advertiser* probably did not exaggerate much when it described the Stave Falls work site as a "model camp":

A visitor to the Stave Falls camp—village would be more appropriate, perhaps—would stand amazed, were he able to take in at a glance the full scope of this undertaking. There are big gangs of men working three or four miles from camp in various directions. In camp itself the sounds of crashing timber, as some huge pine topples and falls, and the fallers drive home the wedge is mingled with the rat-tat of the riveters, the blowing of many whistles and the jar of the steam shovel as she digs her way through the tail-race to the glory hole . . . the excavation immediately beneath the falls.

As the visitor stands on the bridge overlooking the falls, he may get a comprehensive view of the main camp. From this point may be seen carpenters, structural iron workers, woodsmen, mechanics, engineers, concrete and stone finishing gangs and so on, all busily engaged on the various jobs to which they are assigned.

The layout of the campsite described in the article would become familiar to generations of employees and their families at Stave Falls. an upper site close to the dam, where the community centre stood for years, was connected by a bridge to the lower site, which was mostly residential.

On proceeding further down, one will pass in succession the general offices, store and freight sheds, washing and drying houses, bunkhouses and amusement hall, stables, and so on until one crosses the bridge into the "married men's quarters." Here is a model little village. Children may be seen before and after school hours, playing around neat little houses, some of these houses standing on their own lots, other semi-detached, tenement style and so on. About the centre of the village stands the school house, equipped with everything necessary on the inside, but would look much better with a Union Jack flying on the outside . . .

The bunkhouses in this camp are high and spacious and are all fitted with single bunks. This last feature should obtain in any decent camp, but unfortunately it is not the rule in the majority. The "grub" is all that could reasonably be desired, and [we] have yet to see another outfit which includes in its regular bill-of-fare nuts, bananas and other fruit ad lib.

The "grub" was good. At peak construction, the men went through more than three hundred kilograms of beef a week. Every day, the residents of the camp ate 130 dozen eggs, twenty-three kilograms of butter, and one hundred loaves of bread. And as the reporter noted, the company provided a variety of fruit, including oranges, bananas, lemons, apples, pears, peaches, and raspberries.

Heavy rains fell at Stave Falls, resulting in plenty of mud, so it was important to have a "wash house" available. Here the men could have their work clothes cleaned and dried within the day. The provisional powerhouse supplied electricity to heat water for baths and showers. Once the residents were cleaned up, there was plenty for them to do:

Men, women, and children enjoy an evening of dancing at Stave Falls in 1912. Talented musicians and singers were important folk in this small community, entertaining their fellow residents at formal and informal gatherings.

At night, after supper, the men resort to the dance-hall, where they practice dancing and amuse themselves with "sing-songs," when the village belles turn out en masse. *As one sits in the dance-hall and listens to the music, accompanied by the swish of many skirts, it is hard to believe that one is really in a construction camp.*

Those who preferred quieter pursuits had access to a camp library, well stocked by the Provincial Library in Victoria.

Another unique feature of this camp was the presence of a full-time physician, Dr. E.H. Funk, a young English-trained doctor who had recently arrived in Canada. He had at his disposal a small, well-equipped infirmary and a smaller building used for isolation cases, since influenza and other contagious diseases were real threats at the time. Dr. Funk supplied all his own bandages and medications and received $150 a month for his efforts. Workers paid a dollar a month for their own medical coverage; if there were not enough employees to pay the doctor, the company made up the difference. In addition, the company's insurance provider covered damages awarded in any lawsuit up to $1,500. In 1910 it was reported:

Accidents have been rare at the camp and illness almost equally as scarce. A good class of men are now being employed who will take better care of themselves and give better satisfaction to the company. Many of them are taking engineering and other courses from correspondence schools and are striving to better their positions. The company does everything possible to encourage this and give such men every opportunity for advancement.

continued on page 44 >

below:
In this 1911 photograph, these young men stare down the camera with an air of authority that suggests they were accustomed to giving orders. Perhaps they were young members of the engineering staff.

left:
Relaxing in a quiet moment on the construction site in September 1911.

THE STAVE FALLS RAILWAY

Locomotive No. 1 with her crew (left to right), Ross Hunt, Jack Ryan, and Bert Perry, at Stave Falls in 1913. The three steam-driven pistons are clearly visible just ahead of and below the engineer's cab. The crankshaft, running along the bottom, is also visible.

"NOT QUITE AS LONG as the CPR, but just as wide!" was the slogan of the Stave Falls Railway, a ten-kilometre standard-gauge line that linked Stave Falls to the CPR mainline at Ruskin. Although its claim to be the shortest chartered railway in the world may have been an exaggeration, it must have come close.

Western Canada Power built the line in the summer of 1910 to haul building materials and machinery, including some pieces that weighed more than fifty tonnes, to the construction site. The rough and steep wagon road was utterly inadequate. As general manager Robert F. Hayward reported to his superiors in 1911: "Without the railway, the construction work could not have been carried to completion."

Grant Smith and Company, contractors from Spokane, Washington, completed the railway line in less than four months. Motive power was provided by a Shay locomotive, dubbed Number 1, purchased for $10,500 from the defunct Nimpkish Lake Logging Company on Vancouver Island.

The Shay was the leading logging locomotive in British Columbia. Designed by Ephraim Shay and built by the Lima Locomotive Works in Ohio, it was a geared, steam-powered engine that was powerful, flexible, and slow.

The little railway, as well as being essential to build, extend, and maintain the powerhouse and dam, carried revenue freight, passengers, and mail for three decades. Western Canada Power bought a second Shay (Number 7) from the Arrow Lakes Lumber Company and eight log cars (called "skeleton cars") in 1913, when the capacity of the line was increased to haul logs from the Stave Lake operation of the Abernethy and Lougheed Logging Company to a log dump at the mouth of the Stave River. At its peak, the railway carried more than a million board feet a week, much of it high-quality cedar that was milled at Port Hammond. Shingles cut at Henry Stoltze's shingle mill, near Ruskin, also relied on the railway for the short trip to the CPR line. Although the Shay could pull four log cars along most of the line, it could handle only two at "the hump," a particularly steep stretch. Two cars were stored on a siding while the others negotiated the hill. The switching and coupling, high above the Stave canyon, was a hair-raising activity for the crew, but in all its years of activity there were no accidents and only one log was lost.

Passengers travelled between Stave Falls and Ruskin for twenty-five cents, and less if they boarded at one of the half-dozen stops along the line. On some runs they travelled in a comfortable, steam-heated combination passenger and baggage car. On others they rode on benches in the back of a notoriously bumpy truck with flanged wheels, acquired in 1922. The trip could be exciting; there were high trestles along the route that afforded great views of the treetops. Some people simply closed their eyes.

In 1922 the BC Electric Railway Company, proud new owner of both the powerhouse and the railway, electrified the line with an overhead wire. The company retrieved one of its antiquated electric double-ender locomotives, built in 1907, which it renumbered 950, but which was better known by its new nickname, "King Tutankhamun." (The ancient Egyptian king was also brought to light in 1922.) King Tut was soon supplemented by the larger, Baldwin-built Number 980, also a double-ender, which was better for hauling logs, and by Number 1802, a former Vancouver streetcar (dubbed the "Toonerville Trolley" after a popular comic strip of the time), which was used for passenger and mail service. A pair of speeders and additional freight cars filled out the roster.

The construction of the dam at Ruskin not only created Hayward Lake, it flooded a good stretch of the Stave Falls Railway. In 1929 some fifty kilometres of electrified track were rebuilt on higher ground, a feat that required about one thousand metres of trestle.

An extraordinary thing about the Stave Falls Railway was that it used steam, electric, and gasoline motive power, all at the same time. A surviving train order book from 1925 shows the two Shays, the two electric locomotives, the streetcar, and the truck in regular use, typically with more than twenty daily runs, or "orders." A decade later, however, as the slumping economy and competition from trucks reduced the demand for log-hauling, only one to three runs operated each day, mostly by Number 1802 and the truck, with King Tut serving as a backup when 1802 was out of service. The regular crew was reduced to three men, known by everybody at Stave Falls as the "Three Musketeers"—Watt Allan, Bert Kemp, and Charlie Middleton. A Scot from Aberdeen, Charlie began at Stave at the end of World War I as a fireman on the Shays and then operated the truck and the electrics. He and his wife Violet lived for a time in a converted train car on a siding near the roundhouse.

The Stave Falls Railway finally ceased operations in December 1944, after thirty-four years of service. The track was lifted, and most of the rolling stock was consigned to the scrap heap. Number 7 went on to serve three logging companies before being scrapped; Number 1 is believed to be at the bottom of Hayward Lake. Number 980 became a BC Electric snowplough but was scrapped in 1996. A portion of the grade is used as a hiking trail, and a storage shed stands near the Hayward Lake parking lot—the lone survivors of the remarkable Stave Falls Railway.

Clear the way! This photograph was likely taken in the winter of 1911, when more than a metre of snow fell in less than a week and the line had to be ploughed several times a day to keep it open. The photographer, Bert Noble, labelled it simply "Snow at Stave Falls."

Although the townsite was raw in 1911, there was an orderliness to it, with buildings arranged in tidy rows.

This may have been an idealized portrait, because there certainly were accidents. A company letterbook from the time reports that many workers had been on the job for mere hours or days before becoming injured. At one point, it reports that a bucket full of rock accidentally showered the workers below with chunks of stone weighing up to ninety kilograms. Some men were hit or knocked off the forms they were working on.

Safety standards in the early 20th century were minimal. Protective clothing, hardhats, and ear and eye protection were nonexistent. Despite the company's intention to employ excellent workers, at times it had trouble with the employment agency in Vancouver, which sent out poorly trained or unfit men. Still, only four men died during construction, and just one lawsuit resulted. Given the times, that is a rather impressive safety record.

Western Canada Power had a paternalistic attitude toward its construction employees. In a 1911 letter to the Star Realty Company, Hayward (who obviously had more than engineering concerns on his plate) complained to one of its agents:

As I told you when I met you at Ruskin, I am determined not to allow anybody to go into our camp to peddle insurance, mines or real estate, as I know full well that our men have been steadily getting good pay for a number of months, and are easily led away by all kinds of tales and promises that may be put before them.

I say nothing as to the reliability of your firm or the proposition which you are anxious to place before the men; I am absolutely unconcerned as to whether your firm or your proposition is good, but if the men wish to invest their money they can get in with the best of people quite easily, as you well know . . .

I find that during my absence, you had, in spite of advice to the contrary, hung around the camp for several days. Further, in the very face of my advice that you could not go into the camp, you returned that same night.

Under these circumstances, you will not be surprised that I doubt your good faith, and that permission to return to our camp is absolutely refused.

Hayward tried to run a dry camp, presumably for safety reasons. He spent some time tracking down contraband, but booze was smuggled in parcels that came from "relatives," and homemade or bootleg alcohol was readily available.

Despite the amenities provided by the company, for many years the campsite looked bleak. Great swaths of forest in the nearby hills had been clear-cut, the roads turned to mud in the rain, tree stumps from the hastily cleared site dotted the camp, and the mosquitoes were awful. At one point in 1911 swarms of the insects drove clearing crews and survey parties out of the woods for two months. The company solved this problem by using a trick learned during the construction of the Panama Canal: dumping crude oil on the shallow lakes in the area.

The desolate landscape shocked newcomers. Violet Middleton's father, Sydney Salsbury, worked on-site from the early days of construction and stayed on as a powerhouse operator. Violet grew up at Stave Falls and loved the area, but her English mother was not as enamoured. She was quite unprepared for the reality of Stave Falls and had quite a different first impression of the "neat little houses" when she arrived:

When they came out here they had to climb over logs to get around. The places were just like shacks to her. They had tin roofs. Mom said that she cried for days. You can imagine how she felt when she came out here. I think Dad should have waited a little longer until things got a little better up there. But, I guess she got used to Stave Falls. They were together for fifty years!

PROBLEMS ON THE JOB

Construction at Stave Falls was plagued with problems so severe that by the time power started flowing, Western Canada Power's director, Charles H. Cahan, was on the verge of a breakdown in Montreal. General Electric was late in delivering the first generator. The Swiss turbine supplier, Escher-Wyss, delayed shipping its product because the steel runners had to be recast. The excavation of the tailrace was delayed by more than a month because the provincial boiler inspector was slow in issuing a certificate of operation for the steam shovel. Once the shovel was on-site, there was no easy place to dump the rock, so all of it was hoisted out and stockpiled for crushing and pulverizing for use in the concrete.

The concrete was reinforced with steel. These men are wearing hats and nets to ward off the mosquitoes that plagued everyone at Stave Falls in the summer of 1911.

While the dams and powerhouse were rising from the riverbed, the transmission and distribution lines were also being surveyed and built. There were delays here as well, because local municipalities, BC Electric, and the BC Telephone Company all took turns throwing up roadblocks as Western Canada Power attempted to negotiate rights-of-way. The result was that the lines, which should have been finished in September 1910, were not in place to sell power from the provisional power plant, which contributed to cash flow problems.

Another serious concern was the poor quality of the cement (the binding agent in concrete) provided by the Vancouver Cement Company. Large amounts of set concrete had to be removed from the intake dam, and finding a new supplier also held up construction, as detailed in this September 14 letter from Robert F. Hayward to company secretary A.R. Doble:

Owing to various peculiarities in the behaviour of the Portland Cement we have been receiving from the Vancouver Cement Company, we deemed it wise to obtain chemical reports on the cement from very reliable analysts . . . [The most reliable report said that] the cement is of "indifferent quality" and that if it is to be used at all, it should be used in proportions that would be excessive as compared with the best quality of cement. This means, in effect, that the cement we are buying from the Vancouver Cement Company is costing us from 25 to 30 percent more than the price we are paying under the contract.

A rare view through the side of the incomplete powerhouse showing the lower portion of a penstock as it enters the powerhouse to join the turbine.

In the same month, Cahan also wrote to Doble:

We have received less than two thousand barrels of cement although we expected twenty thousand barrels sometime in August under terms of contract with Canadian Cement Company [which owned Vancouver Cement]. We are now delayed by removing large quantities of concrete which proved utterly defective . . . [We] must make arrangements elsewhere immediately for our requirements or practically close down construction. Answer quick.

In November, Hayward wrote the superintendent of construction, Mr. Maxie, asking him to be diligent about testing the concrete:

After all that has been said and done in regard to cement it is of more than ordinary importance that no doubt should be cast upon the quality of our concrete. Besides this, we are undertaking for the first time in the experience of our staff a piece of reinforced concrete construction work and it therefore behoves us to go into the matter with very great care.

Hayward eventually opted for using high-quality White's English Portland Cement for the concrete. But sand, another ingredient in concrete, also proved to be a problem. The sand below the tailrace was found not to be good enough for concrete, so new sand was pulverized on-site from local rock.

These details and others were carefully noted by BC Electric, which had tracked Western Canada Power's activities at Stave Falls for more than ten years. A BC Electric letterbook records the company's interest:

APRIL 20, 1910 [*from Mr. Sperling's diary*]. *Had long conference with Mr. [James] Kennedy, Engineer of* WCP *Co. . . . [who] stated the Stave River Falls plant could not be relied upon to supply more that 19,000 continuous* H.P. *at water wheel shaft and Mr. Hayward was consider-*

ing the advisability of erecting a higher dam in order to be able to make use of the Stave Lake storage to supplement the power output during low water season. Mr. Kennedy also stated that although there were over 100 men employed, no active work was going on at Stave River Falls and [he] did not think they would be in a position to generate power for another 12 mos.

The delays were caused by some serious problems, as the letterbook records:

OCTOBER 1910 [*from Mr. Sperling's file*]. *Conference with Mr. Hunt who reported that the Stave Lake Co. had closed down work for five months and had discharged 500 men on account of their temporary dam having given away, and water was now running over their powerhouse site and had filled up their excavations and foundations with rocks and debris. Mr. Hunt stated that work could not be proceeded with until low water. He also stated that one of the discharged employees had informed him that the Company expected to start generating electricity in Sept. 1911, but that he was of the opinion that, due to the present delay, they would not be in a position to supply current till March 1912.*

Looking up the tailrace, toward the completed powerhouse and, behind it, the intake dam, about 1912. Water was turned into the face of the intake dam on November 5, 1911.

The dam had not "given away"; rather, eleven days of continuous rain had resulted in exceptionally high river levels. Hayward explained: "There was sufficient leakage through the coffer dam to fill the excavation for the powerhouse at a quicker rate than we could possibly pump it out." He did, however, outline a few silver linings in the rain clouds:

The heavy rains coming before we had had time to do the necessary ditching etc. on the railway track has caused us a good deal of trouble for the last fortnight, but as the men could not work on the powerhouse excavation we have put the whole crew on to the railway and it is being cleaned up in good shape and quickly.

The break in the [provisional powerhouse's] turbine runner and the high water together has made a set back of about three weeks in the work of the powerhouse. The rains coming so early has forced us into winter quarters and winter habits, and the men are now broken into working wet or fine and day or night, so that I hope they are now limbered up for the winter.

Other problems, less crucial, were nevertheless annoying. The CPR's passenger service to Ruskin left something to be desired, as this letter from Western Canada Power assistant general manager William McNeill to the CPR's general superintendent indicates:

Yesterday, Friday, I left Vancouver on the 4.20 p.m. train for Ruskin, and reached Ruskin at 6.15 p.m. after a long slow journey, the running time of which would have put to shame a freight train; but then this train may be a sort of a "Glorified" freight train . . . Ruskin will certainly never be developed with such a service as we are getting, and I think that the freight and passenger earnings from that station are such as to warrant a very different order of things.

In July 1911 another pressing matter demanded attention. A letter from the Western Canada Power purchasing agent to the project's explosives supplier detailed a small problem:

A case marked "Hardware" from McLennan, McFeely & Company arrived today, all the sides of which were nailed on. Although it contained 5,000 blasting caps (detonators), strange to say the writer—who opened this case—survives to tell the tale.

Will you be good enough to lay down the law to Messrs. McLennan, McFeely & Company Ltd. that explosives must not be labelled other than what they are, and that wood screws must be used to fasten down the lid of the case—not nails. By doing so, they will reduce to a marked degree the risk of sudden mortality in the personnel of this establishment.

MISSION ACCOMPLISHED: POWER FLOWS

Construction of the first phase of development at Stave Falls was officially completed in January 1912. Stave Lake had been converted into a reservoir and its natural level had been raised by three dams, each blocking one of the lake's three natural outlets. Furthest west was the reinforced concrete intake dam, which contained the penstock openings and their protective gates, twenty-two metres high and forty-nine metres long. The main or sluice dam, also built of concrete, held back the water at the former entrance to the falls. And a few hundred metres to the east stood the 120-metre-long timber-crib Blind Slough dam, replacing a natural barrier formed by ancient beaver dams and piled debris.

Below the intake dam stood the large and impressive powerhouse. Inserted between the walls of the excavated gorge and built of reinforced concrete and steel, it trumpeted the triumph of science over nature. The building measured twenty-seven by thirty metres. Its principal room, illuminated by rows of windows as well as bright electric lights, was the spacious generator hall, where a torrent of water rushed down the two penstocks—steel tubes some 4.4 metres in diameter—into a pair of turbines. The water's weight and speed caused the horizontal shafts in the turbines to rotate so that each produced thirteen thousand horsepower. The turbines were connected to the massive generators, each of which produced 8,825 kilowatts. Two exciters—small generators that provide the current that magnetizes the large generating units—stood at one end of the floor. An array of dials and switches filled a mezzanine above them.

The AC electrical power generated was increased or "stepped up" by transformers and flowed from Stave

facing page:
Riveters join sections of the fifth penstock in April 1925.

Looking east across the Blind Slough dam in February 1924. Designed for flood control, the new dam had fourteen sluice bays; four had hoist operated gates, and ten were closed with "stop-logs"—large, squared pieces of timber that were stacked on top of each other.

Falls to consumers along three 60,000-volt, four 13,000-volt, and one 4,000-volt transmission lines to two substations. The receiving station was fifty kilometres west, at Ardley, in Burnaby. Substations included Powell Street in Vancouver, Sapperton in New Westminster, and Sumas, Washington, where the line connected with the Puget Sound Light and Power Company's system.

The main double-transmission line to Burnaby, the only steel-tower transmission line north of San Francisco, was half again as long as the revolutionary AC line from Niagara Falls to Buffalo, which had been installed sixteen years earlier. In all, Western Canada Power's distribution system totalled almost four hundred kilometres. A year after it opened the Stave Falls plant, Western Canada Power had nearly 1,200 customers and had set 7,658 wooden poles, as well as the steel towers.

More was to come: the plans filed in 1909 called for doubling the capacity of the plant. Intakes already existed in the dam for two more penstocks, and the generating hall had been designed to accommodate two more turbine-generator units. Western Canada Power installed the third unit in 1916, increasing the river's output to about 35,500 horsepower. The company also had the rights to build another power plant just downstream from Stave Falls, at Ruskin.

But Western Canada Power's revenues could not support the costs of expansion. The specifications for the Stave Falls plant had been continually ramped up before the first bucket of concrete was poured. Even as its aggressive marketing campaign promoted power and retail goods, the company was in deep financial trouble. Unable to find enough customers to purchase all its power capacity, it ended up selling a large part of its production to rival BC Electric at a rate that some claimed was below cost.

In retrospect, Western Canada Power might have fared better if the less ambitious plant suggested by Charles Stoess had been built in 1904. That facility came with a relatively modest price tag of only a few hundred thousand dollars, not the $3 million that the first phase of development eventually totalled some eight years later. This smaller plant might have been profitable enough to keep the company competitive and enable it to expand later.

BC ELECTRIC EXPLOITS THE STAVE

BC Electric had long recognized that the Stave Falls plant used only some of the immense potential of the Stave watershed and that it could be further exploited to meet the Lower Mainland's rapidly increasing demand for electricity. In January 1921 the company took control of Western Canada Power and the plant. In the following decade it was able to increase the plant's output from 35,500 horsepower with three units to 87,500 horsepower with five units.

The key to this expansion lay in two ideas: raising the water level of Stave Lake to vastly increase its capacity as a reservoir, and using that water to power a series of generators in its journey to the Fraser. In a carefully co-ordinated, decade-long development spree, BC Electric's engineers raised the height of the dams at Stave Falls, diverted additional water into Stave Lake from nearby Alouette Lake, increased the generating capacity of the Stave Falls powerhouse, and built two new generating stations—one where Alouette Lake water enters Stave Lake, and the other at Ruskin, near where the Stave River empties into the Fraser River.

Improvements to the Stave Falls site began in a near frenzy almost immediately after BC Electric's takeover of Western Canada Power. In October 1921 the fourth generator, which had been stored at the site for some time, was moved onto the main floor. Work began on extending the powerhouse to accommodate it. Mr. J. Meier, Escher-Wyss's erecting engineer, came from Switzerland in the spring of 1922 to assemble and install the turbine at Stave Falls. The unit was working in November, which completed the development that had been proposed by Western Canada Power.

Raising Stave Lake almost seven metres, which amazingly increased its capacity as a reservoir from 171,000 to 471,000 acre feet, was an integral part of BC Electric's scheme. The construction of a new Blind Slough dam began in April 1922. This dam would be higher and would allow excess water to be released—a need made painfully clear after a devastating flood the previous October caused major damage to the powerhouse, railway, and road. A short narrow-gauge railway and an overhead cableway were built to haul construction materials to the Blind Slough dam. A year later the 195-metre-long reinforced concrete structure was complete, with the Dewdney Trunk Road crossing along its

Looking upstream toward the intake dam (left) and the main dam (right) in September 1923. The new intake for the fifth unit is visible in the main dam, and work on raising it is proceeding.

top. The main dam and intake dams were raised 7.6 metres, and a chute was installed to carry some ten thousand shingle bolts (pieces of logs that will be cut or split into shingles) a day through the main dam and downriver to the mill.

Work on the Alouette project started at the same time, with the surveyors going out into the bush to find the best location for the tunnel that would divert water from Alouette Lake to Stave Lake. Water began to flow from one lake to the other in the summer of 1925, and power was first generated in 1928. The additional supply of water made possible the installation of a fifth generating unit at the Stave Falls plant. Engineer E.E. Carpenter puzzled over where to locate it. He examined several alternatives, including placing a vertical turbine on what is now the site of the new powerhouse. This option, which was seriously studied, ultimately proved too expensive, and Carpenter recommended extending the powerhouse westward—requiring a great deal of additional excavation—to accommodate the fifth turbine and generator. He placed the intake a considerable distance away, in the main dam. A huge penstock, about one hundred metres long, six metres in diameter at the dam, and tapering to four metres at the powerhouse, snaked its way under the exciter penstocks and across penstocks 1 to 4 to bring water to the turbine.

The official opening of unit 5 was held on September 19, 1925. Under the watchful eyes of one hundred invited guests, Premier John Oliver took a golden key from BC Electric president George Kidd, inserted it into a special panel in the switchboard, and turned it. This started the small motors in the governor, which admitted a little water into the turbine at first, then more and more, turning the generator faster and faster, until it began to send power to Vancouver. "It was impossible for any words of dedication," reported *The BC Electric Employees' Magazine*, "owing to the noise of the operating machines." Nevertheless Premier Oliver was overheard quipping: "What chance have the Tories now, when I hold the key to power?"

Included in BC Electric's plan to increase power from the Stave was the rebuilding and rewinding of the first four generators to boost their capacity. The last generator went back in service in 1925. The generators produced electric current at 4,400 volts, which was stepped up to the transmission voltage of 60,000 volts (compared with 11,000 volts at Niagara Falls) by a system of transformers and switches. The increased capacity made it necessary to place the new switching equipment and steel transmission towers, nineteen metres high, on a new reinforced concrete deck on the downstream side of the powerhouse. The final stage, the construction of a third power plant at Ruskin, began to produce electricity in 1930, and two more generating units were added at Ruskin over the next twenty years.

Shortly before the Stave Falls powerhouse was replaced by a new facility in 2000, BC Hydro's John Carmichael calculated that because the Stave River plants were so well built in the first place, their value to BC Hydro (successor to BC Electric) has been enormous:

An estimate of the total amount of electrical energy the three plants have produced up until the end of the century can be made by assuming a 50 percent load factor and some guesses as to the start-up dates for each unit. This works out to about 50 million gigawatt hours. That's almost enough to supply all of today's BC Hydro customers for a whole year. To put this into dollar terms and assuming a value of five cents per kilowatt hour, the dollar value of the total plant production would be about $2.5 billion. That's not bad for plants that probably cost only a very small fraction of that.

The engineers, labourers, and technicians who designed, built, and operated the Stave Falls powerhouse ensured that the plant's original structure and equipment produced power for nearly ninety years. The station was capable of significant upgrading, kept the Lower Mainland supplied with electricity, and eventually turned a healthy profit. The system itself is a model of efficiency: water from the Alouette Lake watershed powers three separate plants, and water from the Stave watershed two. This reuse of water power makes sense both environmentally and economically. What more could the builders and operators have asked?

Inside the Alouette powerhouse, about 1928. The 12,500-horsepower vertical-shaft turbine and generator are visible at the left.

Installation of the number 1 Swiss-made Escher-Wyss turbine in summer 1911, showing the curved blades of the two runners and the massive thirty-six-centimetre diameter horizontal runner shaft. The turbine produced thirteen thousand horsepower and operated until the winter of 2000.

TECHNOLOGY AT A CROSSROAD

"HAD THIS PLANT been designed three or four years later, the vertical type of single runner Francis wheel would, without doubt, have been adopted, not only on account of its higher efficiency, but because it would have made possible a very material saving in the cost of the power house." In this post-construction analysis of the Stave Falls hydroelectric plant, Western Canada Power's general manager, Robert F. Hayward, who oversaw the construction and operation of the plant and powerhouse from 1909 to 1920, reflected on how he was hampered by the limitations of turbine technology.

Turbines—sophisticated wheels descended from the water wheels that first allowed people to harness the power of falling water for flourmills and sawmills—are the heart of any hydroelectric generation plant. At hydroelectric plants, water, stored behind a dam, rushes down penstocks (large-diameter pipes) and turns turbine runners (wheels with blades), which then turn gigantic generators that produce electricity.

In 1911 the workers at Stave Falls installed two state-of-the-art 13,000-horsepower horizontal Escher-Wyss reaction turbines, manufactured in Switzerland at a total cost of $93,387. Yet within a decade, these cutting-edge European-made horizontal turbines were considered inefficient and obsolete. The reason, it seems, had to do with the arcane science of fluid dynamics and the theory of lubrication.

Until 1905, the largest and most advanced turbines in North America had vertical rather than horizontal shafts. In that year, three massive 10,000-horsepower Swiss-made turbines were installed at the Niagara Falls plant in Ontario. These turbines, like all vertical-shaft turbines of the time, suffered from the wear and eventual failure of the thrust-bearing, which supported the massive spinning weight of the generator, turbine, and water rushing down the penstocks. To make matters worse, the thrust-bearing on these turbines was inaccessibly located beneath the turbine, which was contained within a large steel and concrete housing in the floor of the powerhouse. Maintenance on the bearing was all too frequent—as often as six times per year—and sometimes required the generator and turbine to be removed with an enormous overhead crane.

The thrust-bearing problem plagued the hydroelectric industry in the first years of the 20th century. It limited the size of the runners and the generating capacity of the turbines. Attempts to bypass the problem included mounting multiple runners on a single shaft in opposite directions to try to increase capacity and counteract the intense downward pressure on the thrust bearing. This solution, employed by the West Kootenay Power and Light Company at its East Bonnington Falls plant in 1906, worked but was expensive and complex. The most widely adopted solution was to mount turbines on a horizontal shaft and thus eliminate the vertical thrust-bearing. This horizontal configuration was used in Shawinigan Falls in Quebec in 1902, and engineers at Stave Falls designed their facility to house the latest in horizontal turbines with two runners.

But horizontal turbines had their own limitations, including the need for a massive and expensive runner shaft, a whopping thirty-six centimetres in diameter, to prevent deflection—the bending of the shaft from the pressure of the water—as well as the need for larger and more expensive powerhouses. As a result, these turbines were effectively limited to 20,000 horsepower.

Soon after Stave Falls was built, a permanent solution was devised to the thrust-bearing problem. Irish professor Osbourne Reynolds's advances in lubrication theory allowed for the creation of a bearing in which oil, under the influence of the bearing rotation, forms a "pressure wedge" that supports the load on the bearing. In other words, there is no metal-to-metal contact, and thus little wear, because the shaft, runner, and water pressure effectively float on a thin film of pressurized oil.

The most successful thrust-bearing to employ the pressure wedge was designed by an American, Albert Kingsbury. The new "Kingsbury bearing," which eliminated the thrust-bearing problem and allowed a return to the vertical shaft turbines, was first installed in Holtwood, Pennsylvania, in June 1912, just as Stave Falls's horizontal turbines were beginning to turn. The bearing at the Holtwood facility, which supports two hundred tonnes, is still in use, showing minimal wear after nine decades. The Kingsbury bearing made its first Canadian appearance at the Kananaskis hydroelectric plant in Alberta in 1913, and the following year at the Cedar Rapids plant in Quebec, which became the engineering standard for Canada.

The Kingsbury bearing virtually eliminated the size restriction—and thus the limits on generating capacity—of vertical turbines, which in turn eliminated the need for numerous, smaller, and ultimately more costly turbines. Simple, efficient, "Francis" wheel single-runner vertical turbines became the norm in North America by 1920. Equally important, the new bearing, which could carry a load of more than five hundred tonnes, was mounted above the generator and the turbine instead of below them. Since the generator and turbine hung from the bearing, maintenance was easy, which reduced the time that turbines were "off-line" and boosted electricity production. The efficiency of these new vertical turbines was impressive, with the massive 55,000-horsepower turbine on the Niagara River reaching a maximum efficiency of 93.8 percent in 1921. Adding to these efficiency gains were important improvements in runner design and the simplification and the refinement of control systems.

When BC Electric raised the Stave Falls dam in the early 1920s and added extra generating capacity, its engineers were limited by the existing powerhouse design. As a result, turbines 3 and 4 were identical to the first two units, and the fifth unit, added in 1925, was a larger horizontal shaft design by a different manufacturer. The engineers did manage to take advantage of current technology by installing a vertical turbine in the Alouette powerhouse and a vertical turbine in the Ruskin facility in 1930.

In 2000 the old Stave Falls powerhouse was decommissioned and a new facility went onstream. The new facility's two highly efficient vertical turbines and sophisticated computer control systems produce nearly twice as much electricity as the five original Stave Falls units did—from the same amount of water!

WYSS & Cie ZURICH

The exciters, in the foreground, provide the direct current that is used to electrify or charge the magnets in the generator's rotor. The magnets must be charged for electricity to be generated. The reflected image of a generator is visible in this photograph, hovering above the exciters.

CHAPTER FOUR

Sawfilers to Superintendents: Operating the Plant

WHEN FLOORMAN PERRY SWAIN worked the night shift on February 5, 1973, he checked the lake level, recorded the air temperature, cleaned a wheel on the fourth unit, and checked the oil levels in the transformers. Routine cleaning and mopping were also done. When Swain was relieved at 7 a.m. the following morning by Bill Fish, he recorded in block capital letters that he was leaving the station in the same state he had found it—NORMAL—completing another phase in a relentless cycle of routine that had begun late in 1911 and continued, with some changes, until 2000.

Keeping the machine running so that power flowed was the goal of everyone who worked at Stave Falls, from the superintendent to the men who cut the grass. Their stories tell us a great deal about what it was like to work day in and day out in a modern, bureaucratized, and technically sophisticated industry.

CONFRONTING THE 20TH CENTURY

"The 20th century belongs to Canada" is how Canada's prime minister, Sir Wilfrid Laurier, saw things in 1904. The builders of the powerhouse at Stave Falls shared Laurier's optimism. Modern industry, organized into factories, using Canada's wealth of natural resources, producing goods in mass, and employing thousands of workers, was the ticket to a prosperous future for labour and capital alike. Electricity, delivered over wires from distant generating sites, would be the driving force of modern industry, and access to cheap, reliable electrical power was considered the key element for business success and economic well-being as well as for domestic comfort.

Throughout the 20th century many electrical utilities evolved from relatively small, local organizations into large systems delivering power to an entire province and integrated into even larger continental grids. The men who managed and worked at Stave Falls directly experienced the development of British Columbia's electrical utility industry. In early-20th-century British Columbia, working in industry usually meant employment in a sawmill or salmon cannery, where raw materials were processed. Although many of these processing businesses were locally owned, by the early part of the century some were part of larger national or multinational organizations that were "highly bureaucratic, heavily capitalized, controlled from outside the region, and run locally by career managers."

The oars have been set aside and Norm Forster is paddling from the bow in this snapshot. Forster joined BC Electric in 1912 and was assistant superintendent at Stave Falls from 1944 to 1954.

Western Canada Power was an example of a national organization: it employed local managers and was connected to important local businessmen, but its professional expertise and money came from outside the province. Similarly, BC Electric, which controlled electrical and street railway services in most of the Lower Mainland and southern Vancouver Island, was created by British capital in 1897, and many issues were referred to London for resolution.

Just as there were different types of companies, there were also divisions within labour. The skilled trades, such as locomotive engineers, electricians, machinists, plumbers, and carpenters, "enjoyed higher status, more job security, and greater income" than the common labourer. Race and gender further divided the workforce and shaped individual and group experiences. Observers of the British Columbia labour situation in the early 20th century noted that nowhere else in Canada was the question of the relationship between labour and management, employee and employer, so much a part of the public debate.

At the centre of this debate were issues of power, control, and "fair" wages. Different companies and unions chose to address these issues differently. BC Electric was renowned for its paternalism, bargaining with the street railway workers union and establishing a profit-sharing scheme in the belief that it had a "duty to look after the men's interests and that in the end it will be to the good of the company." At Stave Falls, most non-management employees belonged to the International Brotherhood of Electrical Workers (IBEW).

Men got their jobs at Stave Falls in a variety of ways. Some came by their permanent jobs by first working on the construction crew. Others had a connection through their family, ethnic, trade, or military networks that opened the door. Flora McIntyre remembers that her dad, a skilled carpenter and boatbuilder, got a job on construction at Alouette courtesy of a friend and fellow Scotsman, engineer Jim Gilbert, who was in charge of construction at the Alouette and Stave Falls dams. When Glen Truax applied to work on the streetcars he failed the eye test and was sent to Carrall Street to see Mr. Vilstrup, BC Electric's chief electrical engineer, whose assistant was an old army friend of Truax's and offered Truax a job as a floorman, although he didn't know "an amp from a volt." Roger Bruder grew up next door to the chair of the union's apprenticeship committee, Al Spencer, who told him: "If you go to grade 12, I'm going to get you a job with BC Electric."

Direct family relationships were important. Doug Forster's dad, Norman, worked at Stave Falls and Ruskin, and Doug put his name in for a job after serving in the air force during World War II:

I would go down and visit some of the people I knew, some of the managers I knew through Ruskin, or from visiting Dad. They all knew me so it didn't really take that long till there was an opening. I got in as a floorman and I worked all over the place. Even at Stave Falls for a while as a relief floorman there, and I was spare operator at Bridge River.

In some families, the tradition of working for BC Electric extended through more than two generations. The Heinrichs were one of those families. Herbert Heinrich started with BC Electric in 1908; he came directly from England to Buntzen Lake, where the superintendent, a friend, had promised a house and a job. Both of his sons, John and Allan (known as Bunt because he was born at Buntzen!), subsequently worked for the company and were posted to Stave Falls at different times in their careers. As students, John Heinrich's sons, Bob and Jack, also worked for BC Electric.

THE JEWEL IN THE CROWN

When power began to flow from Stave Falls in January 1912, the plant and transmission system were shining examples of modern-day technology in a monumental form. The plant was also the biggest power generating facility in western Canada. A few years later, when Western Canada Power merged with the BC Electric, Stave Falls became the jewel in the company's crown. Stave Falls, and Ruskin a few years later, were *the* places to work in the power generating field in British Columbia.

Among employees, there was a hierarchy that extended from the superintendent to the operators to the casual workers digging ditches on the "bullgang." Inside the powerhouse were the operators, described by one old-timer as the "lords and masters," as well as floormen, electricians, and mechanics. Outside the powerhouse the railway had to be kept running, the dams cleared of debris, and men and supplies transported up Stave Lake to the Alouette powerhouse.

In 1916 Western Canada Power had sixty-nine employees in nine departments working to keep power flowing from Stave Falls to customers in Vancouver, the Fraser Valley, and across the line to Bellingham. This did not include men hired specifically to do construction work. The heart of the operation, the powerhouse, required only seven men to keep it running, while out-

Sharing a laugh at the cookshack in about 1912. Bert Noble is in the foreground at the left.

Looking across units 1, 2, and 3 toward the control or switchboard panel. The washroom was to the left of the panel and the offices to the right. Some apprentices became quite proficient at sliding down the railing on the steeply angled staircase at the right; everyone who worked in the plant knew those stairs well.

door maintenance and railway operations occupied twenty-seven men. Head office staff, located in Vancouver, handled accounting, contracting, purchasing, stores, and meter reading. These numbers are quite low, considering that in 1970 there were seven operators and seven floormen at Stave. Western Canada Power may have been operating with a skeleton crew.

Like their contemporaries working for other large industrial firms in Canada, most employees at Stave Falls were skilled or semi-skilled, blue-collar, white males of British or northern European origins, "breadwinners with wives and children who were dependent on them for their economic survival." These workers also formed the backbone of the labour movement in Canada for most of the century. A key dividing point was between the men who earned a salary and those who were paid an hourly wage. In the powerhouse in 1916, all seven of the personnel were salaried, but in the railway department only four of the men were paid a salary; the rest were paid by the hour, from locomotive engineers at forty-five cents per hour to carpenter's helpers at thirty cents per hour. A further dividing point was between those who had permanent or temporary full-time employment and those who were simply "day labourers." Skilled labourers, whether they received an hourly wage or were salaried, typically enjoyed more control over their work than day labourers did.

At Stave Falls the workforce included helpers or floormen, apprentices, and journeymen in a number of trades. For workers, the key to a moderate kind of prosperity was earning a "ticket" and becoming a journeyman in a particular trade. This involved working in the trade for a number of years, first as a helper and later as an apprentice. Some of the trades that worked at Stave Falls—mechanics, carpenters, and locomotive engineers—were long-established when it was built, but others, including electricians, wiremen, and operators, were relatively new areas of expertise.

KEEPING THE POWER FLOWING

In 1916 the work crew in the powerhouse was managed by the superintendent, who had overall responsibility for the powerhouse, the townsite, the receiving station at Ardley, all transmission lines, and distribution lines west of the Pitt River. He was assisted by a chief operator, two operators, two turbine operators, and a machinist. The station was manned twenty-four hours per day, with each day divided into three eight-hour shifts.

Working at a sometimes dangerous and often monotonous job, an operator had to be a "great reader . . . walking back and forth by his meter-panel," monitoring "hundreds of recording devices" from which he could tell that domestic lights were burning and industry's motors turning. Operators also had to be at least legible writers, as well as communicators. There was constant telephone traffic between the operators on shift and staff working on the transmission lines and at the receiving station, and careful daily logs were kept of every communication, weather conditions, water levels, and repairs.

As the system expanded, the station operators were less involved in the details of running the larger system and more concerned with ensuring that the power Stave was expected to deliver was available. One of the critical operations they were responsible for was taking the generators on- and off-line, as required by the demand for power. This job required carefully monitoring the switchboard, communicating with the men on the floor by using simple hand signals (the noise from the machinery made it too loud for anything else), and good eye-hand co-ordination.

Down in the "hole," where the machines were located, floormen and apprentices did the physical labour. John Carmichael, who joined the BC Power Commission in 1961 as a district manager trainee, oversaw the Stave River plants in the 1970s. He had a chance to see the operators in action shortly before the position was eliminated. Carmichael recalled:

It is well known that the operators liked to keep the floormen busy, even if there wasn't real productive work to do, so there was a lot of mopping floors and polishing brass to be done. The operators were hard task masters because they had been there themselves one time . . . it's the old story, where you've got to pay your dues, and the operators made sure the floormen started out right. I think it was a good system, because the floormen wanted to become tradesmen and became good tradesmen. It was good progression for people.

An operator takes down information from a meter, helping to create the meticulous record that was kept of the plant's performance, day by day and year after year. The logbooks at Stave Falls document (among other things) the kilowatt hours generated, maintenance performed, weather conditions, and hours worked.

There was more to the job than polishing the brass. Roger Bruder started as a floorman at Stave Falls in 1965, when he was only twenty years old. He knew a good thing when he saw it; the job offered "good pay and good conditions." Bruder remembers the hourly station checks he and his partner did: "We'd check things like belts, and oil leaks and governors and water flows for the temperature. We'd check all the bearings. And we would go into the switchyard and look at oil levels and check the circuit breakers for air leaks and make sure pressures were up."

SYNCHRONIZATION

Caught in the act: this ghostly image portrays a man near the switchboard at Stave Falls. This picture was likely taken around 1915.

STAVE FALLS MAY HAVE been physically isolated, but from the time it was built, and especially after BC Electric acquired the plant in 1921, it was part of a larger system. Operating that system required finely tuned instruments that could measure demand and control output.

As power demand fluctuated throughout the day, the operators were required to take generators on- and off-line. Every time a generator was started, it had to be synchronized so that all generators at the station were running at the same speed and feeding the same number of volts out through the bus bars and into the transformers on the switchyard deck. As overall demand fluctuated, the operators could decrease or increase, within operational limits, the amount of power being produced by each generator—but they always had to be kept "in tune."

At Stave Falls and other early hydroelectric power plants, synchronization was done manually; operators had to close a breaker on the switchboard at just the right moment to keep everything in synch. It took experience to get the timing down just right. The consequences of not getting it right could be severe. Electricity on the system moves in waves; a rogue or out-of-synch wave would throw the whole system into chaos, causing severe damage to equipment as electrical forces collided.

Floormen and apprentices got more fresh air than the operators did, going outside to operate the spill gates on Blind Slough and the head gates in front of the penstocks. Part of what they were learning was to "start exciters on water and parallel them electrically. How to stop and isolate them to work on them. How to start and stop generators. How to synchronize them and how to switch." Learning usually involves making some mistakes, and Glen Truax, who started as a floorman at Stave Falls in 1945, clearly remembers the first time he was operating: he "took the field off a loaded machine. Big mistake. You could hear it from a long ways away; just a howling because the machine goes up to full speed when you take the field off." Fortunately for Truax, "They didn't hold it against me for too long."

Not everyone who started out as a floorman at Stave Falls became an operator. Some men remained floormen for their entire careers. For many years most operators at Stave had started there as floormen; a man could wait a long time for a full-time operator position to be available. Usually, those who wanted to move up the ladder had to spend a certain amount of time working as temporary relief operators, filling in at summer vacations. Those who were willing and able to move around the system, from Stave to Buntzen to Bridge River, were more likely to eventually work their way into a full-time operator position than were those who wanted to stay put.

A formal apprenticeship program was introduced in the early 1950s. In the 1963 collective agreement between the IBEW and BC Hydro, the two parties agreed to establish an apprentice training committee that would "determine the standards and methods of training and qualifying apprentices. This committee may appoint subordinate committees to prepare and grade the examination of apprentices in each of the several trades."

By 1968 considerable progress had been made on codifying the apprenticeship process with the establishment of selection criteria and regulations governing most aspects of the process, from how courses were to be paid for to the ideal age of apprentices (from eighteen to twenty-five). A fully developed terms of reference for the apprenticeship program appeared in the 1969 agreement and has formed the basis of the program since then. At that time, an apprentice was defined as "an employee in training, and as such, is entitled to be instructed in every aspect of the trade in which he is apprenticed. He is not to be considered as part of normal staff complement except where it is a requirement of station manning."

In the early years, experienced operators and floormen had trained the new men who came in "off the street." Ove Sorley, who started with Western Canada Power on a brush-clearing crew during construction in 1916 and then worked his way through a series of jobs before he became a full-time operator in the 1920s, chose not to rely entirely on his mentors. He writes in his memoirs:

When I first entered the plant as shiftman I began by asking a lot of questions. Some I got answers to and some I didn't. So at the first opportunity I bought a set of Hawkins

At work assembling the stator, the fixed part of the generator containing the copper windings. This picture was probably taken in 1916, when the third unit was installed.

Electrical Guide, also a copy of the American Electrician's Handbook and studied the questions I wanted answers to. After a time I supplied the answers to them.

The differences between the old-style operators and the men who arrived at Stave with a training in electricity as part of a formal apprenticeship program signalled a significant shift in thinking about what made a good operator. Roger Bruder observed the differences between the young men with courses from the Vancouver Vocational Institute and the older men, some of whom had been on the job longer than the new "boys" had been alive. They had extensive operating experience but, with the odd exception, very little theoretical grounding:

The younger fellows [*in the 1960s*] *definitely had this electricity down pat, they understood it quite well. When they were doing the job, they were using their theory basically to get them by; whereas the older fellows would more look towards a step list.*

Ultimately, there was something to be said for both styles of operating. The new breed were more flexible, able to move from plant to plant. The older operators were experts in every detail—and all of the quirks—of the Stave Falls plant.

AUTOMATION

The biggest change to occur at Stave Falls between the completion of plant expansion in 1925 and the time the plant closed in 2000 was automation. This happened first at Ruskin in 1971, and a few years later at Stave Falls. It meant that the plants no longer had to be staffed full-time; instead, a day-shift crew maintained the plant and the rest of the time it was empty. John Carmichael noted that it was hard for the operators "who had grown up in those jobs to see the plant sit there; walk out and say OK, nobody's there."

At Stave, automation was relative. Charlie Scott explains:

They never got so they could synchronize these [*generators*] *from town. They could take them off the line from town. They could open the breaker and let them run free, but then you can only let them run so long without wrecking the bearings. So, they used to have to call people in to shut them down and people in to start them up. They could load them up and unload them, and they could take them off the system, and they could open the circuit breakers and close the breakers. This was sort of a semi-automatic station.*

Operations superintendent Gordon Burr (right) poses with Roger Bruder, who is holding his apprenticeship certificate, for the company newsletter in April 1969. Bruder served part of his apprenticeship at Stave Falls.

After hours, if something went wrong, a member of the service crew (the term used in the 1970s for electricians and operators) would get called by the control operator in Burnaby to go into the plant and fix the problem.

To Roger Bruder the biggest difference between "now" and "then" is the pace of change. For the older men he worked with in the mid-1960s, nothing had changed in the course of their long careers: "A guy who was operating in 1920, in 1965 could have stepped in and done the job just as easily." But Bruder doesn't think that would have been possible in 1985. Computerization had fundamentally altered the job, and operating staff were constantly having to learn new systems. Of course, the men who started working at Stave in the 1910s and probably into the 1920s faced their own challenges. They were working in a new industry, teaching themselves and their colleagues how operate the plants, devising systems, and, at Alouette, operating one of the earliest "automatic" plants on the continent.

Bill Gosbee, who was in charge of electrical maintenance at Stave from 1924 to 1947, was one of the plant's pioneers. He landed a job with BC Electric at Buntzen in 1911. For the first few days he hand-mixed cement, "the worst job I ever tried," but fortune smiled on him and after a few exhausting days he found himself working in the powerhouse as a helper. At Buntzen, Gosbee worked with a group of well-educated engineers, and feeling his own ignorance he decided to pursue his education through the International Correspondence School, following a course in electrical engineering. When he got to Stave:

I found the electrical maintenance was left entirely to me. If we had a burnup or the plant shut down, I got a call through the operator and the rest was up to me. I was only a five-minute walk from the plant. I had to study, and for the next fifteen years I had to really dig to keep up with the changes."

The operators couldn't do their jobs without the work of men like Gosbee, who kept the electrical systems in the plant working, and machinists and mechanics who took care of the mechanical maintenance. If maintenance work was needed on a generator or turbine, the floormen would "isolate and drain," making sure that machine was safe to work on. Then the "mechanical people and general trades people and welders would go in and do the work." The process of isolating a machine and giving a clearance to work on it was serious business. Charlie Scott explains:

All through the powerhouse, instructive messages like this one helped staff with some of the finer points of working at Stave Falls.

If I was giving you a clearance, it would be your prerogative to take that clearance and then make me go through and show you every isolating point, both mechanically and electrically, [to confirm] that that unit is isolated for you to go and work on it. So there's no hydraulic valve that's left open so you could be hurt, or there's no electrical switch left open.

Although the trades had their separate domains, they did overlap. The floormen and apprentices were particularly aware of this, since they carried out the more mundane electrical and mechanical maintenance. One anonymous floorman left this poetic commentary on mechanic John Heinrich's fixation with proper lubrication on a loose slip of paper that was found in a logbook from the 1930s:

Raging floodwaters swallowed up the railway tracks in 1921, providing plenty of work for the bullgang.

There is a man named Johnny Heinrich,
Who, in case of faulty lubrication is sure to find it,
Now this bearing says he has too much oil,
While this one here hasn't got enough,
And again this one here ain't got the right kind of stuff,
Now, after 20 years or so of use,
He investigates tachometers and raises the deuce,
Oil rings are stopped & there ain't enough oil
Literally speaking he makes me boil,
So take heed dear floormen of faulty lubrication,
For Mr. Heinrich patrols the whole damn station.

Just like Charlie Chaplin in the movie *Modern Times,* the men at Stave Falls were part of an industrial system governed by the clock and its noisy counterpart, the plant whistle. Heinrich's son, Bob, remembers that "there was a whistle that would blow at twelve o'clock, there was a whistle at ten to one; and then there was one at one. It was like school, at ten to one that would go off and Dad would have to get out of the house and have ten minutes to get back to the shop." The importance of the whistle wasn't lost on anyone at Stave Falls. Every day for many years, on the way back to the plant from lunch, Heinrich stopped to feed some bread crumbs to Suzy the seagull, who knew the routine just as well as he did.

WORKING OUTSIDE

The crew in the powerhouse was small compared with the brigade of workers who did the general maintenance work and kept the townsite clean, the railway operating, the dams clear, and the lake open. In 1916, twenty-seven people were working outside. This group included skilled tradespeople like Frank Horne, who had his "papers" as a locomotive engineer, as well as carpenters, painters, gardeners, and labourers. These men performed a bewildering array of tasks, some routine, others the product of extraordinary weather or some "higher-up's" decision to replace or improve facilities or equipment. A sampling of work orders from the 1930s includes tasks such as cleaning and painting penstocks 1 and 2, hauling gravel for Tom Welsh's house, salvaging a boat, rolling and watering the tennis court, scraping and painting the switchrack, and painting the washroom, hall, bedrooms, and floors in cottage number 79.

Some of the men stayed with BC Electric for years. Watt Allan, Charlie Middleton, and Bert Kemp, known as the "Three Musketeers," ran the Stave Falls Railway. Middleton evidently got so accustomed to having the rails to guide him that he once went off the road while driving his car because he forgot to steer. Vera Ruddick, Watt Allan's daughter, remembers her dad's daily routine: "The old train went everyday to Ruskin and did all that switching around and bringing things back. In the afternoons he'd just work in the office."

Jim Kelso also stayed with BC Electric for many years, landing a job on site preparation for construction of the Alouette powerhouse in the mid-1920s and then studying for his "ticket" and captaining the thirteen-metre MV *Ruskin* on Stave Lake until his retirement in 1948. Kelso came by his seafaring talents honestly; his father and grandfather had captained "puffers" out of the Isle of Arran on Scotland's western coast, transporting limestone to Ireland. Part of his job, with the help of two crew, was to keep the channel from the dam up to the lake clear so that men and supplies could be ferried to the Alouette powerhouse. He knew Stave Lake and its treacherous drift- and stump-strewn reaches like the back of his hand.

Kelso's daughter, Flora McIntyre, remembers thinking the *Ruskin* was a yacht because "Dad kept it spotless. Took it out every year into drydock, scraped and painted it and varnished it. Spotless." On special occasions such as a visit from the company president, at the request of the company Kelso donned a jacket "with buttons" and a hat that made him look like a seafaring captain.

The seafaring traditions of BC Electric, and later BC Hydro, on Stave Lake didn't end with the retirement of Jim Kelso. Frank Petersen followed in Kelso's wake, running tugboats and supervising a maintenance crew on Stave Lake from the late 1940s until his own retirement in 1981. Like Kelso, he had a healthy respect for the lake, observing that "it can be perfectly calm one minute, and the next the wind will be blowing up a storm."

Charlie Miller, who worked as a "jack of all trades" at Ruskin for many years and later supervised the outdoor crew, started out as a sawfiler. With a family to support, he told the contractor on the Ruskin project, "I'm going to be here every morning putting an application in until you hire me." Charlie stayed with BC Electric until 1962, when he retired at the age of sixty. Family was also on Ken Ruddick's mind when he took a job on the bullgang (maintenance crew) at Stave Falls in the 1940s. He explains that "we were having our first youngster, so I thought I would stick around." Ruddick laughingly remembers that "you'd go down in the morning and as long as you could talk you didn't go to work. It would be ten o'clock in the morning, everybody would be chatting about stories. Then when there was nothing more to talk about you went and did some work."

At the centre of all of the outside activity was "Billy" Matheson, "a born diplomat and glutton for hard work." The foreman of outdoor work for many years, Matheson started with Western Canada Power in 1910 on the construction work as rigging foreman. After serving in the army in World War I, he returned to work for Western Canada Power and later BC Electric. When construction work at Stave Falls was completed in 1925–26, Matheson transferred from the construction department to the operating department, retiring in 1954 as general foreman. In addition to working at Stave Falls, Matheson had a second career as a municipal politician, serving on the Mission District Council for many years and finishing as reeve of Mission in 1955–56.

Captain Jim Kelso on board the MV *Ruskin* at Stave Falls. Built in 1930 at A.L. Linton's Vancouver shipyard, the *Ruskin* was almost thirteen metres long and equipped with a sixty-horsepower Vivian engine.

Most of the people who worked at Stave Falls were men. But occasionally women were employed, not in the trades or on the bullgang, but in domestic jobs. After her husband died in 1917, Jane Hendry was hired to manage three of the camp houses, making sure that the company's inventory of furnishings was properly maintained. In some situations whole families were hired. In 1918 the cook left and the camp manager put "S. Carter and his Wife in the cookhouse." Carter, who had lived in the area since the 1890s, then brought one of his sisters from Silverdale in to help out.

Women typically earned much less than men. In 1947, for example, the relief cook at Stave Falls, a woman, earned 46 cents an hour or $3.75 per eight hours, at a time when labourers were earning 70 cents an hour. One of the relief cooks, Jean Morris, stayed with company for thirteen years, from 1945 to 1958. Sometimes the work women did was informal. John Heinrich's wife, Wally, was trained as a nurse, and although she wasn't on the payroll, in the 1930s and 1940s she was often the first person called when a worker had an accident.

Don Paterson's truck was a familiar sight at Stave Falls.

THE MEN IN CHARGE

Social commentators and pundits watched the utility industry with particular interest. Electricity embodied the modern industrial age, and the form taken by the organizations that delivered it was, and continues to be, considered an important question for everyone in society. Debates raged over whether electricity should be delivered by private or public organizations, and whether those organizations should be small local entities or larger regional bodies. Great interest was also taken in the professionals who managed the utilitites; many were engineers, who were regarded as heroes of the modern age.

Ideas about management have come and gone over the century, but they have always idealized "administrative efficiency" and "improved productivity." Large private enterprises such as the railways also had well-established bureaucracies, building on the British imperial model. Creating and managing the systems that made them work, from power generation to delivery to billing and publicity, required a multitalented workforce and skilled managers.

When Western Canada Power operated the Stave Falls plant, the firm had a relatively small, centralized, and hierarchical organization. Robert F. Hayward, the firm's general manager in Vancouver and an engineer by training, had responsibility for all the company's day-to-day activities. Hayward in turn reported to the company's financial backers, men in Vancouver and eastern Canada. The company was small in comparison with its competition, BC Electric, and with the much larger utilities that emerged in North America. It had a single generating station and two major transmission lines:

one that went south to the United States and another that served the Vancouver market. Stave Falls had a superintendent who answered to Hayward. As general manager, Hayward divided his time between Stave Falls and Vancouver.

After the Stave Falls plant became part of BC Electric, the management structure changed. BC Electric was a larger organization and, as early as 1925, it had a manager at its head office who was responsible for the firm's dams and generating stations in the Lower Mainland—Buntzen, Coquitlam (dam), Stave, and later, Alouette and Ruskin. For many years, the man who filled this position was Fred Elwell, the "superintendent of power plants." Elwell's week consisted of "a trip by boat to Buntzen Lake, another to the Alouette dam by automobile and speeder, another to the Alouette power plant by boat from Stave Falls, with perhaps a trip into Coquitlam dam and a visit to the Stave Falls and Ruskin." At Stave Falls, Elwell's direct subordinate was Tom Welsh; he became superintendent at Stave in 1917 and continued in that position until 1944. Ken Ruddick describes Welsh as "a little guy—his wife was the big one—but he was the nicest guy."

Fred Elwell's retirement in 1940 marked the beginning of a major turnover in senior management at BC Electric. Bill Gosbee remembers what happened: "A new engineer from eastern Canada took over as superintendent of P.H. [powerhouses] who was J.P. Fraser by name . . . Fraser was really sharp. In short time, he changed the various superintendents around." It was likely Fraser's decision, in 1944, to move Tom Welsh to Ruskin and put Bill Henshaw in charge at Stave Falls. Gosbee remembers Fraser's priorities quite clearly: "He wanted the plant kept running and wanted to have questions answered when he got on the phone."

John Carmichael's appointment replacing John Heinrich as manager of the Fraser Valley area, including the plants at Alouette, Stave, and Ruskin in 1970, was another break with tradition. He wasn't an old hand from BC Electric (he had twelve years of utility experience with Ontario Hydro and the BC Power Commission) and he had never been an operator; instead, he had a university degree in electrical engineering. There is an interesting parallel in Carmichael's education forming part of the qualification of management and the increasing emphasis on formal study in the trades. In both cases, formal education, rather than experience, became increasingly important to employers. Bill Gosbee, who retired from BC Electric in 1958, experienced the transition first-hand when, in 1953, he took over as superintendent of laboratory and meter shops. He comments that "there were some graduate engineers under me on this job and I could fairly feel the resentment from them."

Whether the worker had grade 12 and a journeyman's ticket or a degree in engineering, the main responsibility at Stave Falls was the same. In Carmichael's opinion, this was "to make sure that you got all the energy out of the water that was there, because that was valuable stuff. You didn't want to spill water, you wanted to generate as much power as you could." What did change, though, was that "if they didn't create those kilowatt hours, if they went off-line, the customers wouldn't know the difference anyway because the power was available from other sources into that same area. So that was the transition in terms of the importance of those plants in the local area."

PERHAPS YOUR LIGHTS WENT DIM

Occasionally the generators ground to a halt and the all-important kilowatt hours were not generated. Equipment failures, weather, and human error all played a role in making the work at Stave Falls more interesting. On a Tuesday evening in January 1927 the operators at Stave Falls had a spot of trouble, causing lights in the Lower Mainland to go dim. One of the operators later explained that "it wasn't our fault, it was our misfortune. A high-tension switch bushing on our 60,000-volt switch rack failed, and the resulting arc would not break, so we had to drop all our load to break the arc."

Those events were nothing compared with the mayhem that had ensued when the construction circuit grounded in 1924. On that occasion the floorman opened the door to the high-tension room to be greeted

Mrs. Matheson was confident that her husband Bill would make good use of the reclining chair he was presented with at his retirement party in 1954. The party attracted more than 120 people with a program that included cards, dinner, dancing, and the reminiscences of a number of Bill's World War I buddies.

with "a sheet of flame and dense smoke." In the "excitement" the operator "cleared the board," interrupting service to Vancouver and Bellingham. On another occasion in the same year, a series of events led to the switchboard fuses blowing and the operator "losing control of the station." The resulting damage was extensive.

Snow and electricity just don't seem to mix. Usually the consequences are seen in downed transmission lines and power interruptions some distance from the powerhouse. On one occasion in 1955, though, the lines right at the station "took the hit." Glen Truax remembers:

The snow had built up on the 60 kv line and when it fell off, the line snapped up and hit another line. The switch that blew was one of the four 2,300-volt switches just inside the door near the machine shop. The explosion blew about half the oil out of the switch, and it blew open the big folding doors, and it buckled the monorail that they used to move machinery around—you can still see it today. And it blew my lunch bucket right down to the far end of the gallery. There was a German fellow who spent the next year painting to clean up after the explosion and fire that left the powerhouse black with soot from top to bottom. He found a letter of mine from the lunch bucket up on the crane. He worked for a year with a 4-inch brush.

In the early years of operation the management of the transmission system and the powerhouse were integrated, and one superintendent was made responsible for both. Ove Sorley, who later became an operator and ultimately was superintendent at Alouette-Stave-Ruskin, got experience working as a lineman in the winter of 1917–18, when a silver thaw—a cold snap followed by freezing rain—struck the Fraser Valley, knocking out most of Western Canada Power's transmission lines. It took months to restore service. Another "Big Storm" in January 1935 caused a silver thaw that tested the mettle of BC Electric's workers. Eight kilometres of line came down in the central Fraser Valley, and three of the circuits from Stave went "dead." It took five days to restore the lights in Mission, and it was twenty days before full passenger service resumed on the Fraser Valley interurban line.

Rising waters were another hazard to contend with. Meticulous records were kept of weather conditions. Most of the time, maintaining control was fairly straightforward, but occasionally, like "when you got a wet snowfall in October, and then it started to rain on top of that, you had to be pretty careful that you opened up the gates in time to try and start the water running out before it got too high."

At least one flood at Stave Falls caused serious damage to property, threatening the powerhouse and dams. In October 1921 the main sluice dam, located immediately adjacent to the intake dam, still served as the spillway. Leakage from the sluice channel through rock, "which was apparently somewhat shattered during the excavation for the powerhouse site," made it necessary to "timber up one of the south-east doorways of the main floor of the powerhouse to prevent the water which had already found an entrance through this doorway from encroaching further." At Blind Slough, the timber-crib dam and roadway were awash in water and "a large amount of miscellaneous material, such as logs, trees, old floats, gas-boats, debris" was pushed up against the dam. Preparations were made "to dynamite the eastern portion of the dam in order to secure more discharge area for the water coming down from the lake," but the rain finally stopped and the waters began to drop.

The flood waters did considerable damage downstream of the dams, destroying "for a length of some 2,000 feet, two transmission lines, the Ruskin roadway, and about 1,500 feet of the railway track leading from Stave Falls to Ruskin, besides converting a larger section of Mr. MacDonald's farm land into river bed." In the townsite, one house had to be pulled down after its outbuildings were washed into the river. The detailed report on the flood makes no mention of what the people living in the townsite did. (Presumably they headed for high ground!)

One of the factors contributing to the desperate situation was the curious fact that, nine years after the plant came on-line, accurate information was still not available regarding the area of the Stave Lake watershed, its storage capacity at different elevations, the rainfall on the watershed, and the comparative levels of the lake and river for various levels at the dam. Clearly the project's promoters hadn't had the time or money to address a few small details before they built the plant.

ACCIDENTS AND INJURIES

Although logging has always had a reputation as a "widow-maker," working with electricity, especially at high voltages, is also a tremendously dangerous job. Generally, powerhouse operations are safe; line work tends to be more treacherous. In 1912, only a few months after Stave Falls began generating power, one of the company's first linemen was electrocuted on the job:

Mike "Dutch" Grabbe, 32 years old, was electrocuted Sunday afternoon at about 5:30 while employed as a lineman for the Western Canada Power Company on North Road. Twelve thousand volts of electricity passed through his body and there is no doubt but that his death was instantaneous. Details of the accident were meagre, but as near as can be learned he had climbed one of the B.C. Electric poles which are placed near those of the Western Canada Power Company . . . for the purpose of working on the lines of his own company. It is believed that in some way he lost his balance and came in contact with the live wire in that manner. His body was left suspended in the wires until the workman, by throwing a rope about him, jerked him to the ground. It is believed that the death was purely an accident, and that it was in no manner caused by negligence on the part of anyone.

Working on the water, especially on Stave Lake, with its driftwood and multiple logging operations, could also be dangerous:

A sad fatality occurred here on November 24 [1927], which took from us one our most popular and respected companions. Harrison W. Grahame, while at his work removing driftwood from the river in front of the [Stave Falls] dam, and towing it to the catch-boom in [the] upper river, slipped from a boom-stick and was drowned before assistance could reach him. Although dragging operations and the service of a diver were employed for days, and a daily patrol of the water adjacent to the scene of the accident is being maintained, the body has not been recovered.

Most accidents were more mundane than these. Accident reports from the late 1940s and early 1950s list back and knee injuries from lifting, cuts from tools, insect bites, splinters, and broken ribs. Eyes were particularly vulnerable. Scale, cement dust, and, in the most serious case, steel shards, found their way into unprotected eyes. Minor injuries were treated at the site, but more serious cases went to Mission for med-

The railway was electrified in 1922. The crew included motorman Watt Allan (in the cab), trolleyman Charlie Middleton (next to the cab), and conductor Bert Kemp (wearing the vest). The youngster is Bill White.

ical attention. Sometimes a seemingly minor injury could result in serious trouble. In 1935, Margaret Dickinson, the wife of W.J. Dickinson, an employee at the Alouette powerhouse, received a small wound from a nail. Blood poisoning set in and she died, half an hour before the plane that had been brought in to take her to hospital arrived.

British Columbia has had workers' compensation legislation since 1902, and the current system is based on legislation passed in 1917. That legislation recognized "the principle of social justice . . . namely, that industry should bear the burden of its accidents." At BC Electric, safety practices were negotiated as part of the collective agreement between the International Brotherhood of Electrical Workers (IBEW) and the company. As early as the 1920s, workers at Stave Falls were pursuing first aid courses. The importance of accident prevention appears to have received greater emphasis company-wide beginning in the late 1940s. A "Wise Owls" club was formed to promote the use of safety glasses (Charlie Miller was a member), and employees were encouraged to keep track of the number of accident-free days they could collectively sustain. This strategy had some success; by 1951 the Stave Falls crew had gone two years without a single lost-time accident.

The approach taken to accident prevention at that time was rather casual by today's standards. According to Howard Jackman, who joined the bullgang in 1954, safety didn't depend on equipment; "you just looked after yourself and the people you worked with." The systematic use of safety equipment, according to Roger Bruder, began in the mid-1960s, when hardhats and safety boots were introduced. Inside the plant at Stave, the greatest ongoing threat to workers' health was the noise levels. Workers didn't wear hearing protection, and according to Bruder "most of the old operators were deaf." Even for those with hearing protection, it was a hard plant to work in. Roy Taylor, who was there from 1988 to 2000, recalls that it was "noisy, very noisy":

Very hot in the summer. Cold in the winter. It was very hard to actually try and work in there for more than two hours at a time. We were expected to work eight hours with a lunch break, but I think at the most, two hours at a time and have a break, and two hours at a time and have another break was the way most of us worked in there, because the noise was actually unbearable. In the summer when you were working out on the floor with coveralls on and hardhat and hearing protection, the sweat would just be pouring off you.

The noisiest place in the plant was up on the north, or upstream, end of the control panel. Here, "even with ear plugs and hearing protection, the resonation was so loud that it actually hurt. The hardhat usually made it worse because the noise would sort of rattle around inside your hardhat."

TAKEN FOR GRANTED

The users of Stave Falls power have come to take it for granted; electricity is such an inherent part of daily routine that life without it for any significant length of time is hard for most people to imagine. Yet, for the men who worked at Stave Falls there could be no lapse in attention. For the power to flow, every detail of the place had to be attended to; routine work filled most days, but there was always the possibility of a break in that routine, caused by weather or by human error, and workers had to be ready to deal with that. Training was an important part of keeping the machine humming; without it the whole system would grind to a slow halt.

Over time, the daily routine as well as the larger system changed, and workers at Stave Falls had to adapt. Those changes—the introduction of formal apprenticeships and automation—were not isolated events. They affected everyone working for the utility, and they were the product of changes occurring both inside and outside the company. Ultimately, the men who worked at Stave Falls helped to shape, and were shaped by, the bureaucratic and industrial engines that have driven so much of modern life in Canada. Every time they signed-off "station normal," they were saying they had done their part to keep the engines of industry and domestic comfort running—they were modern men, working in what was arguably Canada's most modern industry.

Bert Noble, a floorman at Stave Falls from 1911 to 1917, enlisted in the Royal Flying Corps and learned to fly airplanes. The war ended before Noble had a chance to test his skills in combat.

WAR SERVICE

THE WORKERS AT Stave Falls shared more than a workplace; many of them were also war veterans. A total of 103 Western Canada Power employees joined the armed forces in World War I; fifteen died. For many years, two brass plaques in the powerhouse at Stave Falls listed all of the men who served in that war.

The war left scars. Roy Forster's dad, Norm, fought at Vimy Ridge; Roy remembers that "guns were taboo, he just didn't talk about it [the war], and he didn't want his kids having them [guns]." Some of the stories were bittersweet: Watt Allan returned home with a bride, Violet, whom he had met while convalescing in hospital in England.

During World War II, under the terms of the National Selective Service Regulations, which governed the workforce from 1942 onward, Stave Falls was a "designated establishment": no male employees could leave or be let go without the express consent of a selective service officer. The importance of Stave Falls to the war effort was emphatically communicated to employees by the federal government in a notice informing them: "Your work in this plant is essential to win the war. For your own sake and the sake of your family and friends, help to beat Hitler by regular attendance and the best work you can do."

BC Electric was very concerned about the "vulnerability of our powerhouses to organized attack by informed enemy agents, if any such exist." This concern was addressed in the formation of a special force in June 1940. Ten members of the armed and uniformed 160-strong force were assigned to guard Stave Falls. At the powerhouse, the windows were painted over with black paint to make the structure less visible at night.

The selective service regulations did not prevent the war from touching the lives of the people of Stave Falls. On the home front, staff and their families actively supported the war effort by buying Victory Bonds and doing Red Cross work—knitting socks, hats, and scarves, winding bandages, and sending care packages to those who had enlisted. And at least two men well known to the community—Bill White, the son of carpenter Alf and Annie White, and Sydney Salsbury, whose family had a long association with Stave Falls—were killed in action. Salsbury Lake, one of the lakes east of Stave Lake, is named in memory of Sydney.

55

This class shot was taken at Stave Falls about 1923. The schoolhouse was located south of the tailrace in the lower camp. At right is the teacher, Violet Hunter.

CHAPTER FIVE

My Hometown: Living in a Company Town

On maps and in photographs from the 1950s, BC Electric's compact Stave Falls camp nestles below the powerhouse and main dam on either side of the tailrace. A bridge across the tailrace connects the upper part of the camp, near the powerhouse, to the lower camp, which is located where the recreation site is today. Houses are arranged in tidy rows. Roads and paths connect them to the powerhouse, the recreation hall, the store, and the tennis courts. It is a model town.

The recollections of those who lived there in the early years and photographs from those years tell a different story. The builders of the dam and powerhouse were intent on their primary task—to get the power flowing—and while they made sure the construction camp was physically comfortable, little attention was paid to appearances. In the early years, the whole site was a raw, boulder-strewn construction camp that looked more like the site of a recent hurricane than a model town.

A TYPICAL COMPANY TOWN

For those who lived at Stave Falls, work was never far away. The community was a company town, a "single-resource community." BC Electric owned the houses and property, there was no elected council to govern civic life, and the town's existence was tied to a resource: the water of the Stave River, which was used to generate hydroelectric power. The Stave Falls community resulted from the company's desire to attract and retain a stable workforce in an isolated place. Housing, recreational facilities, and other amenities would help to achieve this.

Company towns have been part of the Canadian landscape since Samuel de Champlain established Port Royal in 1605. In British Columbia, places like Trail and Powell River started out as company towns. As these towns go, Stave Falls was small potatoes. Its residents usually referred to it as "the camp," reflecting its origins as a construction camp and, perhaps, residents' own idea of it as a small and isolated place. One of the things that distinguished Stave Falls from other company towns was that for most of its existence the company that ran it had a monopoly in supplying electricity, so it was unlikely to close as a result of a downturn in the economy.

Although BC Electric had a monopoly on electricity, it didn't have exclusive control over the surrounding land base at Stave Falls, and the community was not completely isolated. Some employees lived in the area before being hired by the company. Others, including the sons and daughters of employees, left the campsite to live in the surrounding area. There were divisions between "them" and "us," those who worked for the company and those who didn't. Similarly, when it came to employment, BC Electric wasn't the only "game in town." But the steady jobs and decent wages made working for the utility better than most of the alternatives.

Beginning in the 1960s, BC Hydro gradually closed the company housing. By the time the obituaries of the men who had begun their working lives at Stave Falls early in the century began appearing in the company newsletter, their hometown was fast disappearing—returning, in the words of one former resident, to "typical bush."

HOUSING AT STAVE FALLS

Many company towns begin their lives as construction camps, and Stave Falls was a classic example. From the early 1900s to 1925 the townsite at Stave Falls housed temporary construction workers as well as the permanent employees charged with keeping the plant run-

Looking across the tailrace toward the upper camp and powerhouse in 1912. A map from the period shows a stable, dining hall, tool house, and machine shop lining the tailrace. Further back were a number of bunkhouses, offices, the amusement hall, and various dwellings, including a number perched on the hillside and visible in this photograph.

ning. The population of the camp fluctuated, depending on the amount of construction underway. Things finally settled down in the camp in 1926, when construction entered its final phase—destruction. Buildings such as the old bunkhouse were dismantled and consigned to the "refining flames." "Sunny Jim," Stave Falls's correspondent for BC Electric's newsletter, kept readers apprised of the changes, writing that "the bakeshop is gone, the dining hall is going, in bits, woodsheds disappear before the advance of electric ranges. Metamorphosis is the order of the day, floors into siding, bunk-house material into roothouse, bakeshop into meat-house and what's left into flame—all tending to improvement and evolution." Some employees also experienced the metamorphosis, transferring from the construction to the operating crew; others left for jobs elsewhere.

In the campsite, just as on the worksite, there was a hierarchy. At the top of the pyramid was the superintendent, who was in charge of both the powerhouse and the campsite. Violet Middleton, who came to Stave Falls as a toddler in 1910 when her father, Sydney Salsbury, got a job as a labourer working on the powerhouse excavation, grew up at Stave Falls and later raised her family there. She remembers: "The superintendent had the big house in the middle of Stave. He was the superintendent of the power plant. But he also had responsibility for all of Stave. Decisions weren't made by everyone . . . We were generally happy with how the place was run."

Flora McIntyre (née Kelso), who was born in 1928, has a keen memory of the camp's geography in the 1930s:

After I was born we moved into a house in the Upper Camp right up by the powerhouse. If you went across the little bridge over the tailrace, with the water coming from the powerhouse through on its way to the Hayward Lake, that meant you were in the lower camp. The lower camp had the bunkhouse and the cook, and the roundhouse there with the trains . . . We had the superintendent living at our end so we considered ourselves the very special of Stave Falls.

A man's position was the most important factor in determining what kind of housing he was provided with, but other factors also came into play, including marital status and whether a couple had children. Ove Sorley explains that when the company built two brand-new brick houses in 1936, "many applications went in for those houses, among them Roy Byrnell's and mine. We were the lucky ones, being the senior operators on the job." Just as the floormen had to put in time polishing brass, they also had to accept what was sent their way in terms of housing. Ove Sorley knew this from personal experience. When, in 1924, he decided as a single man to move out of the staff house because it was too noisy, he wasn't entitled to a house, so he got himself an 8 x 10 tent, which he erected on a platform built of shiplap. This was his home until his bride, Dagmar, joined him. At that point, the Sorleys enlarged the tent and the company wired it so that they could have power and light. The couple finally moved into a house a month before their first child, Milton, was born in 1926.

Not everyone qualified for housing, especially during the construction period. The rules and regulations governing housing at the campsite in 1922 explicitly stated that "the company will provide furnished houses to the power house employees, to linemen resident at

The "lower town" at Stave Falls. The typical home in the Stave Falls camp was built of wood and roofed with tin. Most were relatively small.

IS B. C. CHAMPION

BRUCE GLEIG of Stave Falls, winner of the B. C. amateur boxing championship at New Westminster last month, did some good fighting to get to the top of his section. He had three fights in two days and acquitted himself nobly in each. With a little more experience he will go a long way. His manager, F. H. Horne, declares that Bruce would liked to have gone to Winnipeg for the Canadian championships but lack of finances prevented.

FAMOUS SONS AND DAUGHTERS

above, left:

Triumphant Stave Falls Scottish Dancers pose with their trophies, about 1963–65.

above, right:

Bruce Gleig in his boxing prime. The Stave Falls connection extended to Gleig's manager, Frank H. Horne, who worked on the Stave Falls Railway.

STAVE FALLS MAY HAVE been small, but over the years it produced its share of people who became well known in their particular fields of endeavour. Some were employees; others were the children of employees. One of Stave's famous sons was Bruce Gleig, who came to Canada in the early 1920s from the English Midlands and worked as a logger at Stave Falls. Logging is notoriously hard work, but Gleig was resilient and managed not only to put in a full day in the woods but to train for the boxing ring. His dedication paid off: he won the B.C. Amateur lightweight boxing championship in 1925, sending three opponents to the canvas. Gleig later lived in Chilliwack, where he trained young boxers, sat on town council, and, with his family, operated a cycling shop.

Jack Heinrich, who grew up at various BC Electric campsites, including Ruskin and Stave Falls, donned gloves in a different and perhaps more dangerous ring, provincial politics. Trained as a lawyer, Heinrich was first elected to the provincial legislature as the member for Prince George North in 1979. A member of the Social Credit party, he held various high-profile cabinet posts under Premier Bill Bennett, including labour, municipal affairs, education, and forests.

Cheryle Macdonald, Ken and Vera Ruddick's daughter, shone on a different stage, giving expression to her heritage by establishing a Scottish dancing school at Stave Falls. Macdonald was only sixteen when she started teaching in 1959; the school is still going strong and has toured across Canada and abroad. Over the years, it used the recreation hall at the Stave Falls campsite for recitals, fundraisers, and classes. Years later, dancers still talk about its great sprung floor.

Stave Falls, and to its principal employees in the railway department, but it does not undertake to provide houses for all employees in the railway and construction departments."

Sometimes a house simply wasn't available. When Violet Salsbury married Charlie Middleton in the late 1920s, the couple's first home was a "part of the train, a converted car off on the siding just . . . where the roundhouse was . . . It was very nice. The children used to love to come and visit me because it was just like a big sun room. I had drapes that you pulled across the windows. It really was cute . . . One big room and then a little part where they'd had the smoking section, and then the caboose that was the kitchen. It had a wood stove then." Although the train car was an unusual home, Violet seems to have liked it better than the "tin-roofed shack" that had been her childhood home at Stave Falls. Like the Sorleys, the Middletons got a "proper" house when they started their family.

Dorothy Winders and her siblings, Alice and Frank, lived at Stave Falls in the early 1920s, leaving in 1926. Her father, Frank Horne, came to Stave Falls in 1911, and her mother, Bertha Ramsden, joined him there after their marriage in 1913. War service took the Hornes away from Stave Falls for a time, and when they returned, with two young children, Frank Horne worked as an engineer on the railway. The family lived in the section of the camp known as the flats (where the recreation area is today). Dorothy Winders remembers how the campsite was arranged:

My family lived below the dam and across the bridge on the flats. That's what we used to call it: the flats. There were about ten houses. The people who lived on the flats were the maintenance people—the carpenters—and the railway crew. The people who lived above "on the hill" were the power station families.

As fathers got promotions or left for other jobs, it was possible for those remaining to move into better or bigger housing. Dorothy Winders's family moved three times:

We originally went to the one right next to the store, and then when we moved again we moved right across the street. Finally when the Salsburys moved out we went to Salsburys' bigger house that was right next door.

One of the attractions of living in the camp was the relatively cheap rent. According to Flora McIntyre, the daughter of Jim and Annie Kelso, during the 1930s her parents "rented a house for $15 a month":

That included the electricity and the services of the painters, the carpenters, the plumbers. They didn't do the gardens, but they did the orchards for us, they pruned everything and had it ready so we could go down and eat all the apples and cherries . . . Mr. Tom Welsh [the superintendent] *would order the painters, plumbers and so on when needed. So, $15 a month was a very good deal.*

The provision of free electricity was one of the other perks associated with living in company housing. When the Sorleys moved into their new digs in 1936, they were shocked to find that they had to supply their own fuel.

The camp's geography also had a social dimension, but the weight individuals attached to this varied from person to person. Vera Ruddick, one of Watt and Violet Allan's children, grew up at Stave Falls in the 1930s. She was aware that distinctions existed, but in the end it really didn't matter to her where people lived. Annie

This substantial building served as the camp's hospital during the construction work of the 1920s.

A youthful Dorothy Winders, Bertha and Frank Horne's daughter, smiles for the camera in the early 1920s.

Kelso, Flora's mother, was of the same mind and tackled the situation head on. Flora recalls that she "had a tea party and she invited all the people, up on the hill, on the lower camp and everything, everybody together. My mother didn't worry about other people. She loved everyone and they got along, my mother would see to that."

BEYOND THE CAMPSITE

The Stave Falls community extended beyond the confines of the BC Electric–controlled campsite. It included "outside workers" who didn't live in the campsite but worked for BC Electric as day labourers. They had their own places "in and around Stave Falls." Then there were the loggers, shingle-bolt cutters, the Clarks (who ran the general store) and Ma Simpson (who rented boats), and others who lived up above the dam, along the Dewdney Trunk Road, and up the sideroads, in Stave Gardens, west of Stave Falls.

Although the communities interacted, clear distinctions remained. Especially during the Great Depression, the steady wage paid by BC Electric distinguished its employees from others. Ken Ruddick, who grew up at Webster's Corners in the 1930s and came to Stave Falls in the 1940s, remembers:

There was a differential between the BC [Electric] people and the others because most people in the rural area here were really struggling in those days. It was hard to get work; there was very little money. [BC Electric] people had a steady income coming in and most of them had free electricity and free heat and things.

Another factor that distinguished the two groups was race. Many of the families involved in logging were Japanese Canadians who cut cedar to make shingle bolts, which were then sold to mills to make into shakes and shingles. Most of these families lived along the lakeshore, above the dam, and later in the 1930s, farther up the Stave Lake Road and in logging camps farther up the lake. When the federal government ordered the internment of all Japanese, both Canadian and foreign-born, in 1942, the community was changed forever. Ruddick, who logged in the area for many years, says that after the Japanese were forced to leave, "All of a sudden the lake was just dead."

A big part of living at Stave Falls was the close bonds formed between community members. Here, a group of women and children enjoy tea in the Mathesons' garden around 1927.

Distinctions can cut more than one way, and as far as the people who didn't work for BC Electric were concerned, the entire Stave Falls campsite was "down in the hole." When the construction was at its height, Stave Falls's population was about three hundred (not including children, most women, and Japanese), and about 80 percent of them worked for BC Electric. In the 1930s and 1940s, about 150 "breadwinners" lived at Stave Falls.

IN THE HEART OF NO MAN'S LAND

Gordon Byrnell, who arrived at Stave Falls at the age of eighteen in 1928 to work on the maintenance crew, describes the camp's situation as "in the heart of no man's land." Through most of the winter the roads were impassable, causing a witty correspondent to the employees' newsletter to wonder whether "our company would consider a subsidy to the Pacific Airways." Less patient travellers took the practical approach and made the eight-kilometre trip to Ruskin on the Stave Falls Railway. At Ruskin they could connect with passenger trains on the CPR mainline or a Pacific Stage Lines bus. Sometimes, travellers had the good fortune of riding aboard the GMC truck that BC Electric adapted for use on the rails. It was, Gordon Byrnell recalls, "one of the roughest rides that a person could possibly have because the thing jiggled all the way from Ruskin to Stave Falls. And that is a lot of jiggles."

The miserable state of the roads meant that car ownership wasn't common until after World War II. Vera Ruddick (née Allan) remembers that in the 1930s and early 1940s "there were about four cars in Stave Falls, that's all there was. If you had a car, you were a millionaire. You could look out your front door and see so and so's car going by, so you knew who was going out and who wasn't." One of the "millionaires" of Stave Falls during those years was Norm Forster, who started with BC Electric at Buntzen Lake in 1912 and moved with his family from Point Grey to Stave Falls in 1930. He had a "1928 Pontiac, robin's egg blue. A nice car bought from Bowell MacLean." The car came in useful that

Bert Kemp is at the wheel of the converted truck that was used to transport people to and from the falls on the Stave Falls Railway. It was a ride that was hard to forget!

first year; Forster was commuting to Ruskin, where he worked as a relief operator. The plant there had just opened, and no housing was available for the Forsters until the next year.

Transportation was important because it broke down the social and economic isolation of the camp. By 1936 the Forsters, thanks to the hard work of thousands of unemployed men, could drive their Chevrolet on the newly paved Lougheed Highway from Ruskin right through to Vancouver, where they did their grocery shopping once a month. Other families headed out to Mission. Of course, if you couldn't get to town to shop there were options. At various times there were one or two general stores at Stave Falls. Clark's General Store, located up on the Dewdney Trunk Road, just west of the dam, was a community institution. Other shops came and went. There were also truck vendors, like Fishy Jim, who delivered fruit, vegetables, milk, and even meat.

BC Electric and its predecessor, Western Canada Power, operated stores at Stave Falls. Although company stores have a reputation for ensnaring employees in a cycle of credit and debt, the store at Stave Falls seems not to have been part of this tradition. According to Charlie Miller, "No employee or their families were in any way compelled or pressured to deal at the company store; most of them sent large orders to Spencer's, Woodward's and Hudson's Bay for grocery and dry goods orders and were only charged a minimal freight rate from Ruskin to Stave Falls."

For some people, like Doug Noble's mother, Madeline, a city girl, the isolation of Stave Falls and camp life were very trying. Doug noted that sometimes on payday, which was every two weeks, his parents would go to Vancouver to pick up Bert Noble's paycheque at the BC Electric offices and his mother would have an opportunity to visit family and friends.

As transportation improved, the social isolation of the camp lessened. Bill Gosbee came west from Nova Scotia and landed a job with BC Electric at Buntzen Lake in 1911. He and his wife, Edith, lived at Stave Falls from 1924 to 1947. Gosbee explains succinctly in his memoirs of Stave Falls that as the roads improved, the

Ready to roll: Norm and Marjorie Forster pose with their new Chevrolet in about 1950. This car was not the Forsters' first.

couple made many friends around the Fraser Valley. Charlie and Elsie Scott and their children came to Stave Falls in 1961. Arriving relatively late in the history of the campsite (they made up for that by staying twenty-three years), they found a very different situation from the one that had faced Gordon Byrnell thirty-three years earlier. Brenda Froese (née Scott) thinks her mother, Elsie, "really enjoyed it—the seclusion I don't think bothered her because she could drive and we had a car." The Scotts made regular trips to Maple Ridge, for haircuts on Saturday mornings and to curl on Friday nights at the Granite Club.

The bunkhouse men were less independent of the company. Often single, young, and transient, they paid the company for room and board in the bunkhouse. The cost was deducted from their pay. In 1920 it cost six dollars a month to live in the bunkhouse, with additional charges for meals. A fine line was cut between economy and keeping the men happy. One can imagine the eye-rolling that went on when instructions from the general manager were received by the camp cook in 1920 informing him that "by careful preparation of the meals, not serving too much and yet plenty, by making tasty do-ups of meat and vegetables, great economies can always be effected while at the same time giving the best board."

In the mid-1950s fourteen men lived at the bunkhouse, and more stayed there when a special crew came in. Violet Middleton, who worked as a relief cook in the 1950s, remembers that the men "had a roast almost every day, because in those days, meat was so cheap. It was easy cooking and they really liked things like my banana cream pies, lemon pies. The other cook just made apple and raisin all the time." By 1965 the cook was gone and no more than three men at a time lived in the bunkhouse. Roger Bruder, who started as a floorman at Stave Falls in 1965, explains that "if they had more than three in the bunkhouse the union agreement called for a cook. They just didn't want to pay a cook so there was three of us." Even without having their meals provided, they were "looked after"; the sheets were changed and the kitchen was fully equipped.

IT WAS INCREDIBLE THAT WE SURVIVED

Growing up at Stave Falls offered children plenty of scope for adventure, especially in summer, when whole days could be spent at play. Favourite places included the rocks and pools of the Blind Slough spillway and the old shingle-bolt flume. Up on the lake there were log booms to be explored and logging operations to watch, and the railway was an interesting diversion. These industrial places were not without their dangers. When Annie Kelso realized that her children had had to quickly clamber out of the spillway when the dam was opened, she called the superintendent: "'Mr. Welsh, do you know where the children were playing today? They were down there at the spillway and you opened the gates, and the kids could have been washed down to the Hayward!' . . . He said, 'Now, Mrs. Kelso, I'll tell you what I'll do. I'll have the foreman at the powerhouse blow eight long blasts on the whistle to let the kids know to get out of the spillway.'" Her fears were well founded. In 1930 five-year-old Ruth Matheson drowned, and around 1950 a young boy drowned in Hayward Lake.

Equipped with lights but no helmets, this group of boys is ready to explore the undulating terrain on the rough roads around Stave Falls in the 1940s. Left to right: R. Salsbury, M. Sorley, G. Gegan, and R. Cummings.

The appeal of the shingle-bolt flume cut across generations. Milton Sorley played in it in the 1940s. Charlie Scott's children, who called it the "pig's trough," explored it in the 1960s. When Sorley played there the flume was still in active use, and he was always careful to make sure that "there were no shingle bolts anywhere near." Part of the excitement of being in the U-shaped flume was the small stream of water that leaked through the gate and trickled down the centre. Also, near the bottom, the flume entered a tunnel. Passing through the tunnel tested the courage of youth. Sorley seems to have passed with flying colours, but Jack Scott's imagination held him back. He wondered, "Who knows what's living in there? Who knows who stashed what kind of body parts in there over the years?"

The older children and adolescents ventured further from the camp, fishing on Stave Lake and going for long hikes. Like most rural kids, they found themselves feeling a bit isolated from their high school friends in Mission and looked for ways to get around. By the time he was fourteen, in 1940, if Milton Sorley felt like a treat he would ride his bicycle to Mission or Haney and get a milkshake: "The distance was eleven or twelve miles, the road was gravel. I would take the Trunk Road to Haney and come back by way of the Lougheed Highway and the cut-off past the dam. Or I would go through Steelhead way . . . into Mission and back along to Ruskin and up that way. No gearing, all push." Like Sorley, Jack Scott also got around on his bicycle—for a while. He abandoned the bike when he turned sixteen and came into possession of a 1972 Comet. Then he had to work for insurance and gas money.

Once the rains began in the fall, life at Stave Falls could be a bit confining for children and adults alike. Jack Scott remembers:

When the weather got ugly, you're pretty much stuck around the house. And it was a small house, there was six people living in it. You couldn't turn around without bumping into somebody in the family. The only thing that I think saved my parents' sanity was the basement. It had a finished concrete floor in it and had the old central chimney thing in the middle so you could actually get around the entire basement wall. We probably wore a groove in the concrete roller skating around in the basement.

By the 1960s Jack Scott and his siblings also had two television stations to resort to when things became too boring. Saturday nights were never dull, with *Hockey Night in Canada* and "hamburgers and milkshakes in front of the TV." Another way to kill time was to visit dad at work, just to see the "*huge* building, and *huge* machines—I mean, you could drive a truck around inside it. Where else do you know that there's a building close that you could drive a truck around in?"

Formal, adult-led activities were also a part of childhood at Stave Falls, including those organized around a sport or community service. Flora McIntyre (née Kelso) was taught to play badminton and tennis by her mother and Mrs. Gosbee, and during World War II the girls learnt to knit items for the Red Cross. By the 1960s, as the population of the campsite declined and the local community grew, organized activity was more likely to include a mix of children living in the area. Provided that parents were willing to drive, it was possible, by the 1960s, for children to participate in various activities outside the camp. Jack Scott, who played soccer, credits his parents, like suburban and rural parents in many parts of Canada, with spending a lot of time "driving us, sitting around, watching us, waiting for us to do our thing."

For many years Stave Falls has had a very active Scottish dancing school, which has absorbed the energies of numerous children and adults, including Charlie Scott's daughters, Charlene, Norma, and Brenda. The annual Robbie Burns Day dance was an important fundraiser for the school, as well as being a pretty good party, with Scottish dances followed by polkas and more contemporary tunes.

THAT WAS IN THE DEPRESSION TIME

During the Great Depression, BC Electric did its best to keep everyone employed. But the market for elec-

left:
Everything was uphill from the campsite, including Clark's General Store, located on the Dewdney Trunk Road, just west of the dams. This photograph was taken in 1980.

below:
Swimming with "the gang" on a hot summer day at Stave Falls in the late 1930s. Some kids had so much fun right at home that they didn't want to go away on family vacations.

A youthful Vera and Ken Ruddick. Vera, one of Watt and Violet Allan's daughters, grew up at Stave Falls. Ken came from nearby Webster Corners. They married in the 1940s and stayed in the area.

tricity shrank too, and workers at Stave Falls found themselves working three-day weeks. Keeping a garden was one way to economize. Large gardens, such as the ones kept by the Kelsos and the Allans, were a common feature of many working- and middle-class homes in Canada before World War II, supplying a good portion of the family's fruit and vegetables. Reduced wages heightened the economic importance of activities like gardening for many families.

Although the Depression may have made people more dependent on their gardens, the thrift and economy they practised wasn't new, and it didn't end then. Jack Scott's parents kept chickens, maintained a large garden, and canned fish and vegetables. Gardening was pragmatic; the produce it provided could stretch a food budget, but it also provided a relief from the highly structured workplace.

Another way to economize was to make your own clothes. Violet Middleton sewed most of her and her children's clothes. The extended family and neighbours helped her out:

My mother and sister were working and they used to give me clothes and I would make them over. A lot of clothes that I wore were from my sister. A neighbour across the road had one girl and two boys. And the girl had a beautiful outfit that she grew out of, and she gave it to me for the youngest girl.

Ideally, men were supposed to earn a "family wage," enough to allow women to manage their homes and raise children. For Violet Middleton's mother early in the century, the responsibilities included "keeping the house clean and looking after us, sewing and cooking and boiling greasy old overalls on top of the stove . . . Mum did a lot of knitting and she had a sewing machine too—a treadle Singer—and she made a lot of our clothes." Working for wages, however, was not unheard of. Flora McIntyre's mother, Annie, left the community to earn "extra" money as a nurse so that Flora and her sister could attend high school in Vancouver.

Although the focus was on the home, women engaged in various activities in the community. They organized their own social calendar of tea, bridge, and badminton to alleviate the isolation and tedium of housekeeping, and practical aid was offered to neighbours. Violet Middleton recalls helping some of the younger women, "staying with them before they had their babies and things like that. There was one whose husband was away up the lake to the other plant [Alouette], and she was expecting this baby and she was nervous. When it was kind of due time, I would take her for a long walk and then go and stay overnight."

Perry Cleven, who raised her four children at Ruskin in the 1960s, found herself helping out at Brownies one year. It was, the women of the community joked, a duty that could be escaped only if you "got pregnant or died."

THE SOCIAL CALENDAR

The annual Christmas concerts, summer picnics, and tennis tournaments were important functions on the calendar. The community's social life was orchestrated in part by committees and clubs that integrated social and work life. The company facilitated and controlled these activities by furnishing facilities such as the community hall and providing the land for tennis courts and, for a time, a lawn bowling green.

Typically, management were the titular heads of the various clubs, but staff contributed labour to create and maintain facilities and organize activities. Thus, when the staff approached superintendent Tom H. Welsh in 1922 about moving the tennis court, he informed his manager that he had "looked into this thing carefully and would recommend that if anything is to be done that the new site be chosen, the members will do all the work they can, keeping down the cost as low as possible, such as helping with the grading." Welsh thought the tennis court was important, arguing that "the court as it happens is practically the only diversion here, and as the road is out it's harder still." Welsh was well rewarded for his support, as Stave Falls had an active tennis club for some fifteen years.

A babies' and ladies' party on the porch of the Welshes' house. The Welshes lived at Stave Falls from 1913 until about 1944. Tom Welsh was superintendent for many years.

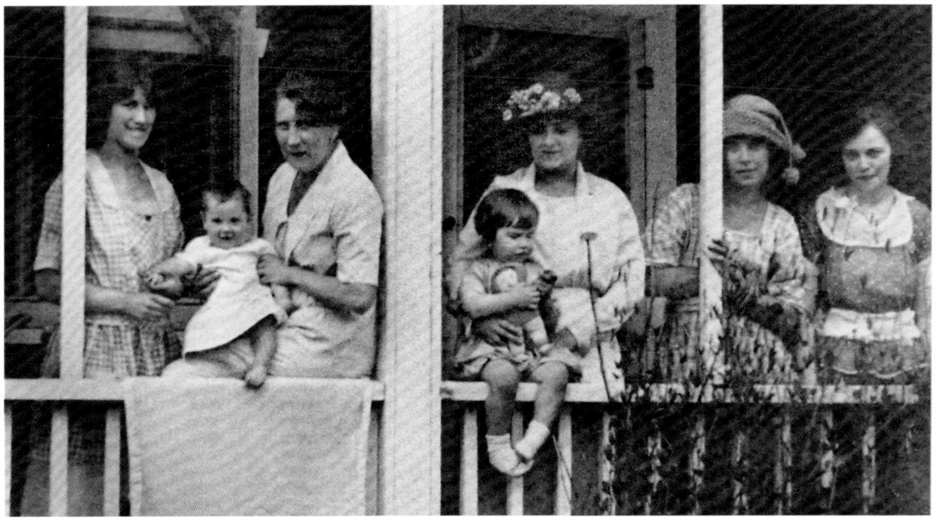

In addition to providing a diversion, competition between camps and with headquarters staff helped to foster a company spirit. The annual contest with BC Electric's Buntzen camp tennis pros was especially hotly contested on both the court and entertainment fronts. When Stave Falls hosted Buntzen in 1927, there was, according to the employees' newsletter account of the event, "something to eat, a little to drink, a few social calls, some sightseeing, eleven sets of hectic tennis, and an abbreviated social dance, and incidents which make conversation, but provide no outstanding contribution to literature, and the day became history."

Off the court, and during the "indoor season," the Stave Falls Social Club convened regular Saturday evening entertainments. The club, under the leadership of that "peppy pair," Ollie Hoover and Teddy Hougham, in 1928 organized "amateur theatricals, whist drives, bridge drives, children's nights" and a large dance, with up to 150 couples attending, once a month. Violet Middleton remembers the social round of the 1930s and 1940s:

We used to have bridge on a Wednesday night and a little whist drive [a game of cards played for money] and dance every Saturday night. Once a month, they had what they called a Big Dance. People would come in from Mission and Whonnock and Ruskin and we'd pay for the orchestra to come. It was waltzes, the foxtrot, very much like Lawrence Welk. The kids came to these dances, so everyone knew how to dance. They would even bring babies and have them sleeping there. The monthly dance was special because it was people coming in, and they were dressed up. The place would be full. They didn't allow liquor in the hall, but sometimes, the fellows would have it out in their car. Women didn't drink in those days much, it was mostly the men.

Tennis anyone? Bert Noble, centre, is flanked by two young ladies. Stave Falls's first tennis court, likely the one pictured here, was located on a sawdust dump in the lower campsite. Not surprisingly, the court tended to sink. It was moved to firmer ground in 1922.

Pulling one of the boys over the line at a game of "pom-pom-pullaway"—a kind of tug-of-war—was a memorable moment for Jacqueline Fraser, one of Charlie Miller's two daughters, at the annual Christmas concert held at Ruskin. At Stave Falls, the Christmas concert was also a memorable affair, with a visit from Santa Claus, a decorated Christmas tree, and entertainment by the school children. Organized by the "ladies," this annual event drew ninety children in 1921. Fundraisers were held during the year to make sure there would be a gift under the tree for everyone. Dorothy Winders remembers the Christmas concerts:

The kids would put on a few little skits and then we'd all wait for Santa to come down the chimney. He'd drop his leg and then it would go up again and we'd go "Oh my goodness what's happened?" Just sitting there waiting— And then he arrived and we were all so happy and we'd get our gift. We didn't know who it was—we just thought it was Santa Claus.

Bert Kemp's annual performance as Santa Claus certainly earned him recognition within the community. Kemp, one of the "Three Musketeers" who ran the railway, was a bachelor and probably the longest-serving resident of the boarding house. For Dorothy Winders he stands out as a "lovable person who always had a chocolate bar for everybody who went to see him."

The tradition of having a Christmas party for employees' children was continued by BC Hydro into the 1960s and 1970s. By this time not enough people were living at Stave Falls to warrant a separate party. Instead, staff and families made the trek into Abbotsford, where they joined with BC Hydro staff. Charlie Scott's son, Jack, remembers:

The Hydro employees would put on a little play presentation. It would take . . . an hour and a half, two hours, and Santa Claus would come in for the kids afterwards, and every kid always got a present. It usually took place in the Abbotsford High School auditorium. That was about the only really major Hydro family function we

ever really attended. The weird thing is, it was always probably about the middle of November, a month before Christmas, no snow on the ground or anything. But hey, Santa Claus was there, let's go!

The summertime equivalent of the Christmas concert was the company picnic. This was typically held on Stave Lake. Flora McIntyre remembers being towed up the lake on a "big barge" with railings around it for company picnics in the late 1930s.

SCHOOLING AND ODD JOBS

The camp had "a little red school house" until about 1924, when a new school was built just west of the intake dam, on the Dewdney Trunk Road. A second school was located further west along the road at Stave Gardens. Grades 1 to 7 (and earlier, grade 8) were taught in these schools. Sometimes the children were separated by age, with the younger children going to one school and the older children going to the other. At one point, there was an attempt to separate the school populations by race—all the white kids going to the Stave Falls School and all the Japanese children attending the school at Stave Gardens—but Vera Ruddick (née Allan) explains that "Mission [School District] didn't like that idea, so we went back the other way."

Like most other rural schools in British Columbia at the time, the quality of the school varied from year to year as young teachers came and went. Some, like Miss Violet Hunter, found their life partners at Stave Falls, while others left as quickly as they came. Doug Noble's mother, Madeline, a trained teacher, was appalled by the quality of instruction in the one-room school at Stave Falls in the early 1930s. The Nobles moved to Maple Ridge after two and a half years at Stave Falls

Occasionally the MV *Ruskin* was put to social rather than work uses. In this photograph, taken around 1940, a group of women is enjoy an outing up the lake.

Almost grown up: teenagers at Stave Falls in the 1940s. This group includes, left to right, Colin Matheson, Flora McIntyre, Walt Lyman, Bill James, Roy Forster, and Alfred Henshaw.

to enable Doug and his sister, Lois, to attend better schools. Doug's father, Bert, commuted to Stave Falls from Maple Ridge for many years.

One story of Stave Falls that will never die involves a teacher who had the misfortune of finding herself in the outhouse when the bullgang came to move it. The legend, as told by Frances Fleming, who taught school at Stave Falls during World War II, goes as follows:

One day the bullgang went to the school to move the outhouse to a new pit. They nailed a two-by-four along the front and back, and lifted the whole outhouse to reposition it on a freshly dug pit. There were crude remarks about it being waterlogged, and much ribald laughter as the edifice was settled on the new spot. The two-by-fours were removed from back and front, and one of the gang announced that it his turn to christen it. As he reached for the door, out sprang the school teacher, red in the face with anger and mortification. She packed her stuff and left camp without as much as a farewell.

Frances Fleming's teaching stint at Stave Falls came while her husband Ray was working at the powerhouse. Qualified as a teacher in 1932, Fleming had graduated into a "depressed world that had wanted no part of me." It wasn't until 1944 that as a married woman with three children, she finally got a chance to teach. That experience rekindled her love of teaching, and although there were bumps ahead—including being promptly fired at the end of that first year because "the boys will be coming home and want their jobs back"—Fleming persevered and had a most successful teaching career.

For the teenagers who lived at Stave Falls, the bumps in the road were more real than figurative. After grade 7 they had to commute by bus to high school in Mission. Until the bus was introduced in the 1920s, parents who wanted their children to go to high school had to find a way to board their children in Mission, or perhaps in Vancouver. Access to high school was especially important for girls because it opened the door to "respectable professions"; before the 1960s more girls than boys in Canada attended high school. Violet Middleton, whose formal education ended after completing grade 8 at Stave Falls, explains:

With my girls, by the time they got to school age they went to school at Stave until seventh grade and then they got a bus taking them to Mission. They went right through high school. One girl, the oldest girl, [*became*] *a hairdresser. The middle one, Jean, she was a nurse. The other one was . . .* [*doing*] *bookkeeping and stenographer* [*work*]. *So they were lucky that they were able to do that. It was because they went to Mission.*

Flora McIntyre went to Magee High School in Vancouver in the 1940s, not because she couldn't get to Mission, but because commuting by bus meant missing out on much of what high school had to offer, including clubs and sports. In the 1970s Jack Scott's experience wasn't too different from Flora McIntyre's:

Once you got to high school, you go, hmm. That's when you really started pining to be able to go into town. These guys always had stories about a party on the weekend. "What?" you'd think. "There was a party on the weekend?" Well, how are you going to things like that when you live ten miles out of town with no vehicle? So all you got to do is hear the stories.

Another interesting challenge facing the average Stave Falls adolescent was how to make a little extra pocket money. Of course some ventures were more profitable than others. Milton Sorley remembers Vera Ruddick organizing the collection of cascara bark. This proved more effort than it seemed worth, in young Sorley's opinion; he recalls that "we were supposed to get a cent a pound, and I guess I got three or four cents' worth—this was dry too! It was an awful lot of work for three or four cents, because this bark was heavy when you took it off the tree, and after you broke it up and dried it, it became very light."

Vera's entrepreneurial talents were not wasted on Sorley, though. He credits her with getting him his paper route, explaining that "she was the *Sun* carrier, and being older she had more worldly contact. I got a bicycle at ten years of age, and I can't remember who had the paper route (*The Province*) with 40 customers before I had it. She steered the agent from *The Province* to me, and I got the route for five years or so. That was a godsend to me because it put money in my pocket." Jack Scott struggled along in the 1970s, cutting lawns and doing "weird jobs like pulling nails out of old lumber" until he finally hit the jackpot at about the age of twelve, landing a job that paid ten dollars an hour at a mill in Ruskin, doing clean-up work.

THE CLOSING OF THE CAMPSITE

One of the reasons companies built and operated towns like Stave Falls was that they hoped the towns would help to attract and hold skilled labour with families at sites that might otherwise be very unattractive to the workers. At Stave Falls, especially until the 1950s, when the "old-timers" started to retire, there was an impressive continuity in the workforce. These retirements, spread over about fifteen years and celebrated in the company of "friends and neighbours" at the recreation hall in the townsite, represented the end of an era at Stave Falls. Gone were the men who had started their careers with Western Canada Power, seen service in World War I, returned to work for BC Electric, and weathered the Great Depression, when "no one moved, or died, or was fired." Retirement meant leaving Stave Falls, since the homes there were reserved for active staff. Some retirees simply moved up the road, while others left the area entirely.

The 1962 collective agreement between the International Brotherhood of Electrical Workers (IBEW) and BC Electric states that the company would "rent housing for married personnel at isolated areas" and at other locations "as will be determined mutually from time to time." The 1965 agreement lists the isolated sites where BC Hydro "will continue to provide rental housing for operating married personnel on the regular staff." The eight sites on the list included Bridge River, Cheakamus, and Clowhom Falls, but not Stave Falls. That meant that while BC Hydro might choose to continue to provide housing at Stave, it was not bound by the union agreement to do so. As families left Stave Falls, their houses were closed up and eventually demolished.

By the time Vera Ruddick had a paper route, she and her family were living up above the campsite in their own house on the Dewdney Trunk Road across from Clark's store.

The last operations employee to live at Stave Falls was Charlie Scott, who lived there from 1961 to 1984. The brick house he occupied, near the Hayward Lake picnic area, is now the park manager's home.

The closure of the townsite was a product of a number of factors, some internal and others external. At BC Hydro, from the perspective of both management and labour, the townsite was not working well. John Carmichael, who was responsible for operations at Stave, Ruskin, Alouette, and other Fraser Valley plants in the early 1970s, explains why:

Employees were not free to paint the bedroom and tear stuff apart and do things and feel that they had some ownership, and the employer had to use union contracts, union wages for everything that went on in that building. It was part of the jurisdiction of the IBEW, *so it became expensive to do fairly simple things that a normal homeowner would do for themselves . . . I think in general Hydro found that the townsite was not working well from the employees' point of view, nor from the employer's point of view.*

The expense of operating the townsite was one factor contributing to the decision to close it. Another was changing ideas about the appropriate relationship between employer and employee; the paternalism that fostered company towns and allowed them to function was out of fashion. Wartime liberalization of the laws governing unionization and the postwar economic boom strengthened the position of labour. Workers felt less vulnerable and more able to take control over their lives. The experience of the Depression also had an impact. People were less trusting of authority, and there was a general move toward a more egalitarian society.

This florid float, the creation of employees and residents, won third prize in Mission's May Day parade in 1936.

From a practical perspective, there really was no reason to continue to operate a company town at Stave Falls. By the 1950s it was possible to commute to and from the site by car. Most employees settled nearby, in Mission or Maple Ridge, but a few commuted all the way from Vancouver. One factor that made moving out easier was the rapid growth of the surrounding communities, especially Maple Ridge. Employees had more housing choices than ever before, and housing generally was more affordable. One-third of Canadians could afford their own home in 1948; by 1961 this proportion had risen to an astonishing 60 percent. Living in town had other advantages. Children were closer to larger schools, and families could have a wider and more diverse social circle.

Technology also played a role in reducing the workforce and changing the organization of work at Stave Falls and other small company towns in postwar British Columbia. Along the coast, for example, cannery communities were closed as refrigeration technologies made it possible to consolidate operations in a few larger communities. At Stave Falls, technological changes resulted in the partial automation of the plant, shifting control away from operators at the powerhouse to system controllers at Burnaby Mountain and reducing the number of people needed to run the plant.

Visitors to Stave Falls today would be hard-pressed to believe that it was once a community of up to three hundred people, but for the many people who lived there the memories are very real. Some loved it. They have nothing but positive things to say about living in a company town; they feel lucky to have lived there, cherishing the bonds of friendship and sense of family. Others didn't like it, for a variety of reasons. Their stories help us to better understand both Stave Falls and a fascinating piece of Canadian history—the company town.

THE GREAT OUTDOORS

FOR THE OUTDOORSMAN, Stave Falls was an ideal setting. In 1926, the *BC Electric Employees Magazine* described the adventures of "Messrs. Kamb, Ryall, and Sobie" on their annual trip into the "wilds of the Upper Stave." Here they found "high water, a ten-mile hike, lots of grief, very little game, but oodles of hungry fish." Getting up the lake was always part of the adventure, and there are many stories of trips that didn't quite go as planned. Ove Sorley, who purchased a 16-foot Peterborough canoe in the 1920s to go fishing up the river, recalls one trip that ended with both him and his fishing partner, Frank Lisk, cold and miserable after an accidental upset.

The Stave's best-known fisherman was Charlie Miller. Raised on the banks of the river, near Ruskin, Miller loved the Stave Valley. Hunting and fishing were much more than simply a diversion for him. He wrote: "As I grew up, I spent three-quarters of my life on and adjacent to the Stave River and its nearby feeder lakes. My two main occupations were cutting firewood . . . and fishing and exploring every nearby lake and stream, catching and keeping just as many fish as we could use without wasting."

Miller was generous with his knowledge, teaching others how to fly-fish and hunt. Doug Forster, who grew up at Ruskin and Stave Falls, benefited from Charlie's expertise, explaining that "he was our fishing mentor" and crediting Miller with teaching "us what we know about fly-fishing." Whether you were an avid sportsman or simply joined in the occasional jaunt up the lake for a picnic, the ready access to beautiful and wild country was one of the pleasures of living at Stave Falls.

left:
Pleased with his catch, Charlie Miller poses for the camera after a day's fishing on the upper Stave River in 1948.

centre:
This is no joke; we caught them! Hazen Nickerson (right) and Fraser Stanford pose with trout they caught in the Stave River.

right:
Enjoying a camping expedition on the Stave River in 1912.

A bird's-eye view of the new powerhouse (right) and the old one (left) in February 2000. Also visible are the intake and main dams, above the old powerhouse, and, if you squint, the Blind Slough dam (far right). The bare patch above the new powerhouse is the site of the new switchyard.

CHAPTER SIX

Entering the 21st Century: Stave Falls Comes of Age

After World War II large-scale hydroelectric development was considered key to British Columbia's economic growth. The provincial government, through the British Columbia Power Commission, created in 1945, worked to bring electric power to "rural" British Columbia. Under Premier W.A.C. Bennett, it eagerly promoted projects on the Peace and Columbia Rivers. But Bennett's plans for the Peace were frustrated by BC Electric's control of the large Lower Mainland market, so the premier solved the problem by nationalizing the company in 1961—"capitalism by the people," it was called. In 1962 the province combined BC Electric and the BC Power Commission into a single entity, the British Columbia Hydro and Power Authority. BC Hydro, as the authority came to be known, was given responsibility for developing first the Peace River project and subsequently the large power projects on the Columbia River.

Between 1960, when BC Electric completed its Bridge River project, and 1984, when power started to flow from the powerhouse on the Columbia River near Revelstoke, the amount of hydroelectric power generated in British Columbia mushroomed. The Bridge River project went back to the late 1920s; shelved during the Great Depression, it was reactivated in 1945. When the last of the Bridge generators came on-line in 1960, "its 512,000 kilowatt output would account for half of BC Electric's

The beach on Hayward Lake, below the powerhouse, opened in 1983. Two houses on the site were part of the Stave Falls townsite and are now used by park staff.

total capacity and be the biggest power project in Western Canada." Stave Falls, by comparison, pumped out a measly 52,800 kilowatts when it was running at full tilt. It had become a small player in the power game.

FINDING A PLACE FOR STAVE FALLS

By the early 1980s BC Hydro was facing the difficult decision of what to do with the aging Stave Falls plant. Not only was its power no longer essential to keep the lights on in the Lower Mainland, but the plant was exhausted. In more than one report it was described as "tired, obsolete, and inefficient," or words to that effect. BC Hydro had three basic options: it could abandon the site, rehabilitate the existing plant, or redevelop the site and build a new powerhouse.

The question of what to do with Stave Falls wasn't entirely new: as early as 1959, just as the Bridge River project was being completed, BC Electric had reviewed the role of the whole Alouette-Stave-Ruskin complex in its system and made a preliminary assessment of what would be involved in redeveloping the site. No action seems to have followed, and the question of what to do with Stave lay dormant until 1969, when another study was commissioned. This study concluded that by 1974 Stave would be "obsolete and considered to have no value."

What did have value in 1969 was the site; the engineers who wrote the report for BC Hydro concluded that if the reservoir and dam were raised, a two-unit, 216-megawatt plant could be installed at Stave to provide "dependable peaking capacity." Three options were explored; the preferred one involved building a new plant and dam at Stave Falls and adding a unit at Ruskin, but not significantly changing the water levels in Stave Lake. It would minimize the effects of redevelopment on the area's property and industries.

RECREATION

An interesting aspect of the 1969 report is its discussion of recreational opportunities on the lake. Lowering the lake during construction would provide an opportunity to clear dead standing trees, making it "suitable for recreational use as well as increasing property values above the shoreline." Although recreation was recognized as a benefit, it was still considered a frill. The report's authors did not think the cost of improving the lake should be included in calculations of the cost of redevelopment: it was not a "true cost of redevelopment."

The report recommended improving the reservoir because "the intensive use of existing lakes in the Lower Mainland suggests that substantial benefits would accrue as a result of lake improvement." Historically, BC Electric had never been very keen on the use of its reservoirs and other properties for public recreation. John Carmichael, who managed Alouette-Ruskin-Stave in the early 1970s, remembers that the old-timers on staff "stood guard over the company properties." As a private business, BC Electric was very focused on production, and as far as it was concerned its plants and reservoirs were private property. The creation of BC Hydro changed things. With ultimate ownership resting in the provincial government, it was harder to argue that the utility's properties were private. As a consequence, BC Hydro began to take initiatives to facilitate and manage public access.

BC Hydro's initiatives were not taken in isolation. By the late 1960s there was immense pressure on the provincial government to improve public access to Crown lands generally. In the political rhetoric of the day, public lands had become private preserves, with forest companies gating publicly financed roads and dictating who could and couldn't travel on them. Although most of the public attention focused on access to Crown lands, BC Hydro also came under fire for its (or its predecessors') failure to clear reservoirs, rendering them unappealing or downright dangerous to recreational users. Environmental awareness was also increasing. Sprawling clear-cuts and debris-strewn reservoirs came to symbolize a cavalier attitude toward the environment.

In the early 1970s BC Hydro was already spending money on clearing Williston Lake, the reservoir created by the huge W.A.C. Bennett Dam on the Peace River. John Carmichael remembers that the leadership on this came from head office staff, who were "sensitive to the need to respond to the public concern over reservoirs." They had the support of the chairman, Gordon M. Shrum. Carmichael remembers thinking that if BC Hydro had the money to spend on Williston, "then they should be able to find the resources to make Stave Lake more accessible." He points out that "right next to Vancouver they were sort of getting a black eye for turning people away from a much more available resource that could have gotten much more positive press."

The pressure from the public for access wasn't an abstract issue. Carmichael recalls:

You had to go and chase the guy with a canoe off the lake, you had to use your staff and your people as police and say "Get out of here." Then you get into the argument where the guy says, "I'm not hurting anything, what am I doing?" So you had look at it from their eyes and [admit] . . . it's not that he's doing anything wrong; although there's always the safety issue.

That Stave and Hayward Lakes had strong recreational potential was no surprise to people living in the area. Jack Scott, who grew up at Stave Falls in the 1960s and 1970s, recollects that "it was quite a popular campsite for people coming out from the town. Even to this day, you get lots of people from Vancouver coming out and spending an afternoon at the lake."

Along with rapid population growth in the surrounding area—Mission's population increased from 2,668 in 1951 to more than ten thousand by 1971—an ever-increasing number of people were looking for places to play. Without any amenities, the results were sometimes a bit messy. Scott notes that there weren't any outhouse facilities, "so when you were trekking through the bush you had to watch where you were going. Watch out for

A walking trail has been developed along the old railway right-of-way on the shores of Hayward Lake.

that telltale white flag!" There were other things to watch for too. Jack Scott remembers that "Mom and Dad didn't like us going all the way to the Blind Slough because that's where the rowdies hung out."

CLEARING "SLAVE LAKE"

The decision to clear Stave Lake was made in 1970: "BC Hydro's concern for recreational and aesthetic values in British Columbia was demonstrated by an undertaking to clear Stave Lake, 35 miles east of Vancouver, of unsightly trees and debris, and ultimately to create a public recreation area for fishing, boating, and picnicking." The real problem, and the reason recreational boaters were turned away, was that the debris-strewn lake was dangerous to small boats.

Some clearing of the reservoir had been done during construction; surviving records tally up the board feet of cedar and hemlock cut along the river and at low-lying points on the lake. The pace of clearing, though, was slow. In a 1911 report on construction, Robert F. Hayward noted that the land to be flooded "is all more or less heavily timbered and will eventually have to be cleared." He proposed to proceed with clearing "as the demand for power requires more storage capacity." The cost was of concern to Hayward: "There are large areas where the net cost after deducting value of timber will be about $100.00 per acre."

Hayward's report does not include any discussion of aesthetics or recreation, nor does it address the question of whether clearing the reservoir would have a positive or negative effect on the operation of the power plant. Instead, and perhaps not too surprisingly, given the precarious state of Western Canada Power's finances, Hayward focused on the short-term costs of logging in the reservoir, noting that "the clearing is being done as slowly as possible in order to take time to arrange for the sale of the timber. At present the company has on hand over $10,000 worth of logs."

It seems that much of the clearing that did take place prior to flooding was selective and restricted to the most accessible areas. The scope of the job facing BC Hydro and its two partners, the BC Forest Service and Correctional Services, in 1971 was daunting. According to an engineering report, up to four thousand hectares of drowned timber (snags) and high-cut stumps had to be removed. In addition, there were several thousand hectares of floating debris, some trapped behind snags and stumps. Also, large volumes of free-floating debris roamed the lake at the whim of wind and water levels, and huge quantities of water-logged and sunken debris lay on the bottom of the shallow waters. During drawdowns, when the lake level was lowered, this debris was exposed.

The tedious work of pulling stumps or cutting them close to the ground, piling the debris, and burning it, was performed by inmates of a number of corrections facilities, who nicknamed the operation "Slave Lake." One of their camps was located six kilometres north of the dam. The inmates had managed, by 1980, to clear 360 hectares. At this pace—forty hectares a year—the lake would not be completely cleared for one hundred years. This timeline was considered a bit too long, and a review of the project in 1980 produced a number of pragmatic suggestions, including focusing efforts on relatively easily cleared areas that would produce "immediate benefits to navigation and recreation." The advice seems to have been heeded. The pace of work picked up in the early 1980s, with almost five hundred hectares cleared in 1982 alone. By 1986 the lake was judged safe enough for the inmates to graduate from stump-hauling to building a parking lot and boat launch, making public access easier.

facing page:
Deputy Warden J.M. McHale poses proudly with material cleared from Stave Lake. Much of the work clearing flooded timber from the lake was done by inmates from nearby corrections camps, including young offenders who learned to make shakes, fence posts, and other products using the timber.

IF YOU CAN'T BEAT THEM...

By the 1980s it was clear that the increasing number of sunseekers and boaters on both Hayward and Stave Lakes required active planning and management. The construction of the boat launch on Stave Lake formed one part of BC Hydro's strategy to direct public activity into areas that were away from high-risk zones around the dams and powerhouse. One longstanding risk at Stave was from spilling (releasing water) at the Blind Slough dam. Spilling caused the water to rise rapidly below the dam—an area popular with sunbathers and

The latest fashion in work attire: striped pants and a hardhat, modelled inside the powerhouse in 1972.

swimmers—and threatened to drown anyone who couldn't move faster than the rising water.

Area residents and BC Hydro employees understood the dangers. Vera Ruddick, who grew up at Stave Falls, remembers that "as a kid, you knew that those gates weren't safe; you just didn't go near them." As the area became more accessible and the population grew, managers couldn't count on users knowing about, or appreciating, the risks. Over the years, and in spite of various warning systems and efforts to limit spilling to nighttime hours, there were drownings at Blind Slough. To address this problem, a new recreation area opened in the early 1980s on the west shore of Hayward Lake, just below the powerhouse tailrace. This area included a beach and parking facilities, drawing people away from the dangerous Blind Slough spillway.

When the park opened, Jack Scott's family was still living in a company house at the site. He remembers the trade-offs. Although recreational facilities were improved, Hydro limited access to other areas. Scott notes:

It's changed so much. You can't access a lot of the areas now because Hydro has them gated off or chained off or fenced off . . .The public kind of screwed themselves by abusing the area. People would be getting killed in the bottom end of the Blind Slough because they're down there when [Hydro's] starting to spill water. Eventually the public outcry is, "Oh, these people are dying because Hydro isn't doing enough to secure the area." Well, if they secure the area, you're not going to have a nice place to go to. It's kind of a Catch-22 situation.

There were advantages to living in a park. Scott remembers that "you'd come home and they'd lock the gate and you wouldn't have to worry about anybody driving down there any more."

MAKING—AND CHANGING—PLANS

While progress was made on the recreational front through the 1970s and early 1980s, work related to planning for the future of the dams and powerhouse seems to have stalled. The likely explanation for this is that in the 1970s, BC Hydro was preoccupied with the much larger projects on the Peace and Columbia Rivers. It wasn't until 1981, twelve years after the 1969 report was filed, that the question of what to do with Stave Falls was revisited in another report. By then, the turbines and generators at Stave Falls required considerable maintenance and repairs. The strategy adopted was to perform only enough repairs to keep the plant in operation until a decision was made about its future.

The 1981 report concluded that BC Hydro could economically rehabilitate or redevelop Stave Falls. Further studies followed. Three years later, in 1984, a detailed review was conducted of a number of redevelopment and rehabilitation options, as well as of the more radical step of abandonment. This detailed review concluded that the best option for Stave Falls was rehabilitation. An important factor in this decision was that the power from Revelstoke had just come on-line and as a result BC Hydro had a significant power surplus on its hands. At Stave, the first order of business in the resulting long list of tasks was improving the dams—securing them so that they would be able to handle the "probable maximum flood," as well as to withstand earthquakes. This work was carried out in 1985–86.

While rehabilitation work was proceeding, major changes were occurring in both BC Hydro and the utility industry in Canada and the United States. Until the 1980s, "electrical utilities in industrialized countries focused almost exclusively on the construction and operation of ever-larger generation facilities." Since then, for a variety of reasons, including stricter environmental controls and deregulation, utilities have looked for a wider range of ways to supply electricity. In 1989 BC Hydro introduced two programs—Power Smart and Resource Smart—that reflected the shift away from megaprojects. Resource Smart "aims at improving the efficiency of existing generating facilities or the installation of generation equipment at hydro storage facilities." The decision to rehabilitate Stave Falls was revisited in 1989 in the context of this new policy. The reviewers found that redevelopment would "provide significantly higher net economic benefits, energy efficiency, and reliability than rehabilitation would."

Jim Hinds (left), mechanic, and subforeman mechanic Ron Davis (centre right), discuss the rehabilitation work at the plant with BC Hydro president Gerald McGavin (centre left) and Don Swoboda, vice-president, electrical operations, at the open house held at Stave Falls in 1987 to celebrate seventy-five years of operation. The event attracted more than five hundred visitors.

One of the more interesting aspects of the work done at this time is the "environmental credit" given

Inside the new powerhouse, showing the huge Kaplan turbine runner, shaped like a ship's propeller. The worker is standing on one of the adjustable blades.

under the Resource Smart program to projects that were considered to have little or no long-term environmental impact. Stave Falls qualified for the 15 percent credit because there would be no new environmental disturbance. The credit, intended to recognize that there is an environmental cost associated with new construction, applied equally to the rehabilitation or redevelopment options at Stave.

The redevelopment option was selected for other reasons. First, there was the issue of the condition of the existing facilities, which were "inefficient (with only 70 percent overall efficiency); do not comply with current fire, worker safety, environmental or design standards; and have advanced in age to the point that maintenance requirements have become excessive and continued service is unreliable."

Then there was the question of how much could actually be achieved by rehabilitation. At the heart of the problem were the old Escher-Wyss turbines. Nothing could be done about their low efficiency, since it resulted from their design and construction. The fifth unit, installed in 1925, was in worse shape than the older machines. As a precautionary measure, patterns were made of the blades in case they failed in service and new ones had to be fabricated quickly.

Even with rehabilitation the turbines could only be expected to last another thirty years, and they could not be made to produce any more energy. This last point was important. Resource Smart emphasized efficiency, and by 1990 the water resource at Stave definitely was not being used efficiently. When everything was running smoothly the plant could produce 52.5 megawatts; a new plant could produce up to 120 megawatts. Another factor that made redevelopment more appealing than rehabilitation was that, by putting the new powerhouse on a different site, the old plant could be operated until the new one was ready to roll.

A question often overlooked in the reports and discussions about redeveloping or rehabilitating the Stave Falls plant is why it needs to be operated at all. With so much power flowing from the Peace and Columbia Rivers, who needs Stave Falls? Bob Ellis, the on-site construction manager for the new powerhouse, acknowledges that as "a single identity, it doesn't really make a big blip on the screen." But it is one of a number of small plants in the system, and taken together they do make a difference.

The plant's location near the Lower Mainland is also considered important. Don Swoboda, who retired in 1998 as senior vice-president of power supply, explains why proximity matters:

Failure of the long-distance transmission is one element of it. It is also important from a point of view of voltage support; at the end of a long transmission line you have to control the voltage to bring it down, because the voltage will go up in unloaded long transmission lines. So you have the voltage control functions of these local plants close to the load . . . These close to the load plants are also particularly important to restore service in event of a collapse of the modern BC Hydro electrical system. They provide that backup with the capability to get those plants started and the supply re-established in the Lower Mainland, while you are trying to build the main supply from the north. To some degree that supply of generation upon loss of northern transmission is fed by the connections to the United States. When you lose your northern generation, most of the energy comes in initially from the south, but close proximity of generation to the load is an advantage: the transmission is shorter, there is less outage exposure, and it's probably more reliable in re-establishing the system plant by plant.

JUST A LITTLE PEANUT

Assigned to the Stave Falls redevelopment project shortly after he joined BC Hydro in the early 1990s, civil engineer John Boots describes it, in comparison with Revelstoke, as "just a little peanut." Stave may have been small, but "it had every aspect of engineering that Revelstoke had, just on a smaller scale . . . You still have to do as much design work in a small structure as a large one."

This particular peanut had a number of key components, including "a new intake structure at the east end of the existing main dam, two power tunnels, a two-unit, ninety-megawatt surface powerhouse, an excavated tailrace channel, a tailrace berm, and a new switchyard located on a terrace on the eastern bank."

The design of each part received intensive scrutiny, with various options considered. For example, the option of using penstocks rather than tunnels was studied; and consideration was given to actually building a new dam, rather than rehabilitating the existing structure. The type and capacity of the turbines received a lot of study early in the game.

Two types of turbines were considered: vertical Francis and vertical Kaplan. The Francis-type were cheaper by about $4 million, but the Kaplans had other advantages. The flow of water through the latter can be controlled at the wicket gates and by adjusting the blade angle. This means that the turbine can operate at higher efficiency over a wider range of conditions, from ten to forty-five megawatts. This was important because Stave, on occasion, must be operated at quite low loads to facilitate the management of water levels in the salmon spawning grounds below the Ruskin dam. As a result of these advantages, the Kaplans were chosen.

Each of the generators at Stave Falls can produce up to forty-five megawatts, for a total plant capacity of ninety megawatts. How much power is actually produced depends on the demand for electricity and, of course, the availability of enough water to generate the power. Deciding on the capacity of the plant required a careful study of demand, water supply, and other factors.

Before work could begin on the construction of the new powerhouse, BC Hydro had to pull together all its planning studies and apply to the provincial government for an energy project certificate. Randy Bourne, who was the project manager for construction, remembers that there were "lots of meetings, lots of late nights, weekends, trying to get this through." He sees the process of preparing the application as crucial to the project's success:

The most time we spent mulling over was in the initial application for an energy project certificate. That's right up front in the project. When you apply to the government, which we did in November of 1994 . . . [you must supply] your preliminary engineering work, your project justification, what public consultation you had done to that point of time, environmental impacts and the mitigation and compensation plan, how you were going to contract and build the project, as well as providing estimates on the labour force.

The filming of Paramount Pictures's *We're No Angels* at Stave Falls in 1989 involved the use of props large and small, including the mannequins shown here. One of the more memorable constructions was the "bustling" community that was created along the shores of Stave Lake.

Although he was trained as an engineer, Bourne doesn't think that the engineering component was the most challenging:

Being an engineer, you are trained to look at the engineering aspects because that is what actually produces the power which we sell, but there are lots of experts in those fields and it is a fairly well-known technology. Sure, each site is different, you have to adapt things and there are certain challenges with each site, but that is not the real challenge . . . The real challenge is dealing with the expectations of the local people, the local stakeholders, the local governments, the Ministry of Environment, the Department of Fisheries and Oceans, and other interest groups.

The public consultation process started in 1993 with an open house at Stave Falls. The turnout surprised BC Hydro: one thousand people were expected, and four thousand showed up. Clearly, the community was very interested in the project. Much of that interest was focused on the environment and fisheries, but there was also interest in work opportunities, the effect of the project on existing electrical services, the impact of construction on local traffic, and transportation issues more generally. That 1993 open house was just the beginning of a process of continuing communication and consultation. For Randy Bourne, the public consultation process "certainly opened my eyes to people's values—what they see as valuable and of interest to them. It always amazed me how people have such diverse views on similar issues. It was interesting."

The preparation of a water use plan was one of the requirements of the energy project certificate. Here, a BC Hydro biologist collects information about kokanee living in the upper Stave River in 1995 for use in the plan.

The consultation process didn't end when the provincial government gave BC Hydro the go-ahead to begin construction. In fact, the terms of the energy project certificate required BC Hydro to complete a number of management plans and studies relating to fisheries, wildlife, heritage, recreation, and water use. These plans and studies addressed issues raised by the public and First Nations, and their completion required more consultation. During construction there was also continuing communication with the community to identify and reduce the noise, traffic, and other disruptions caused by the work.

YEARS OF DRILLING AND BLASTING

Once the energy project certificate was granted in June 1995, BC Hydro was ready to move ahead quickly with actual construction. Bob Ellis, the construction manager, describes the stages of work and his responsibilities:

All the on-site construction, which is the drilling and blasting of the tunnels; drilling and blasting of the area for the new powerhouse; blasting out the intake structure; building the coffer dam on Stave Lake and at Hayward Lake, to keep Stave Lake and Hayward Lake out of the worksite; and then all the new construction involved in installation of the new equipment, right up to commissioning and running the plant.

The decision to use tunnels rather than penstocks to conduct water from Stave Lake to the powerhouse required drilling and blasting that went on over the course of three years. One of the trickier aspects was the underwater work. John Boots explains how this was done: "We left a rock plug that stuck up above low water, and we were able to excavate and blast from that rock plug. In fact, we drilled from the top surface down and from the sides and blew the plug in one shot. It created quite a wave." In the spirit of recycling, excavated

Examining the health of fish stocks formed part of the water use planning process. Here, Wesley Antone, a member of the Kwantlen First Nation, cuts a tissue sample for mercury testing on Stave Lake in August 1998.

rock was piled up against the downstream face of the main dam to reinforce it against earthquakes. The rock also widened the dam and made it possible to increase the width of the Dewdney Trunk Road across the crest from one to two lanes. That pleased local residents.

Once the tunnels had been excavated, they were lined with "shotcrete"—concrete blown onto a mesh of reinforcing steel, which created a smooth interior surface. Lining the tunnels was done at Stave for two reasons. First, there is less chance of loose rocks breaking off and being swept through the turbines, where they can get lodged and cause damage. Also, the lined surface is smoother than the rough rock left by blasting, so the water doesn't lose energy to friction as it passes through. Ultimately, this results in the generation of more electricity.

Despite all the planning, once work started there were some surprises. During excavation for the new intake, it was discovered that the eastern end of the main dam rested on a boulder rather than on solid bedrock. Excavations also revealed that the foundation rock for the coffer dam—the temporary dam that would keep the worksite dry—was riddled with seams of sand. Since water can seep through sand, the seams had to be sealed with shotcrete. Keeping the site dry couldn't be taken for granted. John Boots explains:

There was a lot of work that had to be constructed when the reservoirs were held down at a certain elevation and so there were some real challenges in trying to get the work done without getting flooded. In one case, we did have a small flood of our coffer dam for the intake during a time when inflows into the reservoir were high. We had to detour the road and do a little remedial work, but it was fairly minor. That was a challenge, getting portions of the work done when the reservoir was low.

The need to stay dry also had to be reconciled with operational requirements for water at Stave and downstream, at Ruskin. The sheer size of the hole excavated for the powerhouse—descending four storeys below the level of the lake—is impressive.

Thousands of people—both BC Hydro personnel and private contractors—worked on construction. Bob Ellis recalls:

Daylight penetrates one of the new powerhouse intake tunnels. In the spring of 1996, A.L. Sims and Son Ltd. won the contract to excavate two tunnels to carry water from the reservoir to the turbines.

At the peak we had about 180 people on-site. It went up and down. There were all different phases, so you're talking thousands of people in total. People building roads, then they are gone. You get people cutting down trees and then they're gone. You've got people drilling and blasting and then they're finished. After that, the concrete crews come in. So there's a lot of different people on-site. Then you're down to electricians and the painters.

Added to the on-site personnel were the BC Hydro staff members based at the company's offices in Burnaby. To encourage communication and team spirit, all the staff assigned to the project from a wide range of disciplines—engineers, environmental experts, public consultation specialists, project managers, economists, support staff—were located on a single floor. Randy Bourne considered building a "high-performing work team" as part of his job, so he looked for ways to foster "the serious business of having fun," through golf tournaments, billiard tournaments, coffee parties, and lunches. Cash rewards were given to individuals whose ideas saved the project money. The idea, in Bourne's words, was that "if you do something that is going to save us money, or improve the schedule, or do something that will improve the design, there will be a reward for this." All the teamwork paid off. The first unit came into service in October 1999, and the second unit followed in December. The complex project was completed on time and under budget.

A WINDOW ON THE PAST

Physical evidence cropped up regularly to remind the construction crew on the new powerhouse that the site had a history. Bob Ellis remembers that when clearing began, the crew unearthed foundations, garbage dumps, and even an old tree fort. Paul Rapp, the resident engineer in charge of geological and soil issues, remembers exploring the area where the new switchyard stands. It was dense forest with lots of deadfall, but "everywhere you went it became clear that this whole area was very active." When the road and powerhouse site were excavated, horseshoes, bridle bits, and some wood stave pipe were found. Even the remnants of the temporary powerhouse built by Western Canada Power in 1909–10 were uncovered.

The history of the site wasn't simply an interesting diversion for the workers. The construction records, especially the drawings and photographs produced by Western Canada Power and BC Electric, were essential in understanding how the dams had been constructed originally. Some of the photographs show the original Stave Falls. Located just downstream from the main dam, the distinctive rock that jutted out into the river was still there when construction began on the new powerhouse. The coffer dam built to hold back Hayward Lake was located where that rock sat. When it was removed, the rock was excavated to allow the water to flow through the new tailrace into the lake.

Randy Bourne liked to stand on the rock and remember the picture "where the water is pouring over the rock in 1909." It gave him a chance to reflect on where the river used to flow. Rather than consign the excavated rock to the rubble heap, Bourne had part of it cut up and made into commemorative paperweights. He remembers: "Everyone wanted one these paperweights, with a plaque on them—'Stave Falls'—as a little memento of the project."

Not all of the reminders of Stave's long history as a power station were excavated, buried, or otherwise obscured by second growth. The imposing 1912 powerhouse was still standing. One issue that BC Hydro had to address was what to do with it. Colin Gurnsey, manager of land and social issues at BC Hydro, commissioned a study in 1994 to "look at whether or not the old facility once made surplus could in fact be utilized in some way." He thought that, since it would cost money to tear the building down and restore the site, it was worth looking at what that would cost, and what else—such as a visitor centre that told the story of Stave Falls and hydroelectric generation—might be done with the site and the demolition money.

The study confirmed the economic and cultural benefits of reusing the building. Asking these kinds of questions was an important step. Gurnsey acknowledges that it wouldn't have been possible without the experience gained and changes in the corporate culture that had occurred in the 1970s and 1980s, as BC Hydro began operating the parks and making its reservoirs and facilities accessible to the public.

Charlie Scott, retired electrical serviceman, and electrician Alex MacLellan, leaning against an old friend, the generator of unit 2, at the official shutdown of the original Stave Falls Plant in the winter of 2000.

Gurnsey's interest in history and archaeology goes back to his childhood in Victoria, where he "remembers people like Mungo Martin and others giving talks to school kids about Kwakiutl art" at the BC Provincial Museum. He recalls that at about eight years of age he got "really interested in Egyptian hieroglyphics and tried to learn how to translate hieroglyphics from the *Encyclopaedia Britannica*." At BC Hydro, Gurnsey's job has included responsibility for the company's archaeological work, and he has an interest in corporate and technological history. This personal fascination was one factor in promoting the idea of reusing the powerhouse. So too was the interest of others in the company, who saw developing the powerhouse as a visitor centre as a way to create economic opportunities around the generating stations.

Support for the idea waxed and waned between 1994 and 1998. At one point Gurnsey thought the most realistic plan might be to simply "put some signs up and allow people to walk into the building and look over the railing . . . and that would be the extent of it." Finally, in 1998, the creation of a visitor centre in the soon-to-be

WATER USE PLANNING

Fishing downstream of Ruskin on the Stave River in the late 1990s.

SPORTS FISHING HAS BEEN popular on the bars below the Ruskin dam and powerhouse on the lower Stave for many years. Fly-fishers are drawn to the spot by the cut-throat trout that lurk there. The gravel bars are also important habitat for spawning chum salmon. Over the years, BC Hydro has worked with the Department of Fisheries and Oceans to accommodate fishing and the spawning salmon. In the last fifteen years or so, intensive efforts have been made to work with stakeholders, such as DFO, to improve habitat, manage flows, and make fishing safer for those who do so below the dam. The results of the collaboration at Ruskin have been gratifying.

The issues at Ruskin reflect larger questions related to water use that can be summed up with a simple question: Whose water is it? Is the water there for Hydro to generate power, or does it belong to recreational users, plants, fish, and animals? What about the loggers and other industrial users? How about First Nations? All of the Stave River facilities lie within the traditional territory of the Kwantlen First Nation.

When BC Hydro received its energy project certificate in 1995, giving it the green light to proceed with the redevelopment of the site at Stave Falls, the provincial government required that it address these questions in water use plans that direct how the Alouette and Stave watersheds will be managed. Although the intensively consultative process had examined the river from many perspectives, the interests represented have always been there; they are an integral part of the natural and human story of the river. The end result is a new, and relatively complex, definition of "station normal."

decommissioned powerhouse at Stave Falls was approved in principle by the board. Critics might argue that a visitor centre is a frill and that BC Hydro should focus on producing electricity. To Gurnsey, that misses the point:

It is really important if people are going to conserve energy to understand the mechanics of what has to go on to get that energy to them. If you turn on your switch and the light bulb comes on, if you are aware that people have to work, people have to maintain equipment, water has to be spilled—which has an impact in some instances on, say, fish or aquatic habitat—you realize that all these things are for your benefit, but need to be managed wisely. So, if you don't have any vision of that—in most cases you can't get that vision—where can you go to understand how these things happen? If you can get that vision, you might in fact use that resource wisely and treat it as an asset or a good that is not to be squandered. That is really what I would like to see happen.

With the board's approval in hand, it was time to do detailed planning. This required not only thinking about what would go on inside the visitor centre, but also how the building could best be adapted for public use without destroying its historic fabric. That meant pulling together a wide range of skills from within and outside Hydro, including designers, historians, communications specialists, conservation experts, architects, and engineers. Getting such a diverse group to work together has been, in Gurnsey's opinion, very interesting: "The engineering guys work basically on rules—it's sort of like absolutes—there are co-efficiencies that they have to follow in this system and this procedure in this system and so on. That is a disciplined process, whereas the creative side doesn't have the same discipline. I would say that's where I have been losing my hair."

John Boots made the transition from working on construction scheduling for the new powerhouse to being the project manager for the visitor centre. He explains how he got the job:

I got a call from the manager of project management, and one of his fellows went on vacation, broke his leg, and had to have it re-broken and all pinned. He was off for a couple of months and was assigned to Colin Gurnsey as a project engineer on this project. Then, he decided he didn't need to work anymore—so I was asked to take the job on.

For someone more accustomed to working on new construction, this job has been, in Boots's words, "like renovating an old house. You want to modify some of the wiring, you find that this panel has been added onto over here, that one's been added on over there, and you're chasing wires all over the place trying to figure out what the hell is going where."

Tracking wiring has been the least of Boots's troubles. Some of the cost-cutting measures taken by Western Canada Power have come back to haunt him. Boots observes that "cement must have been at a premium because they would go to great lengths to leave voids wherever they could in the concrete, or to just throw rubble and rock in voids to reduce the amount of concrete poured." Those economies were not reflected in the drawings and have made the cost of getting the building ready for public use higher than anticipated. But not all the surprises have been unpleasant: when a

Getting the old powerhouse ready for visitors. The metal casing on the first turbine has been partially removed to facilitate viewing. The visitor centre opened in the fall of 2001.

square vent cover was removed from just below the roof on the eastern wall, it turned out that it was covering a round, two-metre-wide opening. A window has been installed in the opening, providing more natural light within the powerhouse.

COMPARING THE OLD AND THE NEW

With the old and new powerhouses so close together, it's inevitable that they will be compared. John Boots enjoys looking at the arches inside the old powerhouse "just above the units on the north wall." He thinks, though, that they "must have taken forever to form." According to Boots, the new powerhouse doesn't have any arches because "the cost to build an arch versus just a rectangle is too high, so we try not to build things to curves."

Roy Taylor was a sub-foreman electrician at the old powerhouse at Stave Falls during its last years of operation. In his opinion, the equipment in the new plant isn't likely to last as long as what was installed in old plant. He explains how the newer equipment is different:

The windings are usually a lot lighter; the material they use [*for the turbine casing*] *is completely different to the big old cast iron—everything is . . . made to finer limits. And I don't think even the people who built Stave Falls envisioned it to be lasting ninety years; but everything was overbuilt in the old days.*

facing page:
The exterior switchyard and deck were added on the downstream side of the powerhouse in 1924 to accommodate increased production. Most of the towers were removed in 2001 to open up the deck for visitor activities.

One thing that really stands out in the minds of people who worked in the old plant is the fact that they could see all the machinery—the penstocks, exciters, governors, turbines, and generators. Everything was exposed, and the operation of the machinery was relatively straightforward. Comparing the old equipment with the new, Roy Taylor comments that with the former, "what you saw was what you got." Today, "all the modern stuff is shrouded in cabinets and electronics—you don't really know what is going on inside."

In the new powerhouse, there is very little to see. The plant is quiet, so it is hard to understand that water is moving through the building. For those who worked in the old plant, its "simplicity" is part of what made it a good place for apprentices to learn, and this is why it should make a good place for members of the public to learn about hydroelectricity. Charlie Scott explains that the turbines and other equipment at Stave are "wide open":

You can see everything. If you go over to the new station, you can go down into one little room and you can see the shaft of the generator spinning around; everything else is covered—you can't see anything. Here, you can see the wheels spinning around right out in the wide open. You can see and explain how the governor works. When we take that cap [*the metal casing on the turbine*] *off, you'll be able to see where the water comes down, how it went through the wicket gate, how it hits the wheel, how it drops down into the draft tube. It's just an ideal situation for learning.*

A DYNAMIC VISION

Not everyone has shared the same dream for the Stave: the lake, river, and valley mean different things to different people. This book has focused on the hydroelectric story at Stave Falls. For the men who surveyed the area and built the powerhouse, the water of the Stave River was meant to be harnessed, its energy captured in turbines and transformed into electricity. They were a mixed lot: capitalists and accountants, engineers and skilled craftsmen, labourers and cooks. The recent redevelopment of the site shows that the dream of these early developers was a dynamic one, capable of changing with the times.

It is easy to focus on the construction story, with its many twists and turns, but the longer story is that of the operating power plant. Working at an early hydroelectric facility and living in a company town is a fascinating part of the history of 20th-century Canada. That much of what happened in the course of operating the Stave Falls plant was "normal" is no surprise, but neither should it be taken for granted. It represented the hard work, over many years, of everyone at the plant, from sawfilers to superintendents. It has also meant that, for almost a century, the people and businesses of the Lower Mainland have been able to rely on a stable and secure source of hydroelectric power.

Chronology

This chronology summarizes the development of hydroelectric power at Stave Falls.

1899 The Stave Lake Power Company is formed.

1900 Charles A. Stoess carries out survey work for the company.

1903 A log house is built to accommodate surveyors, and a suspension bridge is built across the falls.

1905 John Hendry becomes president of the Stave Lake Power Company. Site clearing and preparation work undertaken.

1906 Road from Ruskin to Stave Falls is improved and the campsite expanded. William Kennedy Jr. of Montreal is hired to do design work.

1907 Up to two hundred workers erect the piers of the sluice dam on the main channel of the Stave River.

1908 Lack of funds halts most work at the site.

1909 Stave Lake Power Company is sold to Western Canada Power. Coffer dam is built across the west channel of the river to allow for excavation of foundations of the intake dam. A contract is signed with Escher-Wyss to supply the turbines. Robert F. Hayward is appointed general manager.

1910 Work begins on clearing Blind Slough dam site. Provisional powerhouse completed. Tenders let for various components of the project. Contract is signed for excavation of powerhouse tailrace. Sluice dam is completed and construction of the intake dam begins. Railway is completed in September; 467 employees are working at Stave Falls in December. Work is set back three weeks by the failure of one of the turbine runners in the provisional power house and by high water in the fall.

1911 Machinery and electrical equipment begin arriving at the site by train in April. The powerhouse foundation is completed in June and work begins on erecting the steel frame. A large pothole under the toe of the intake dam is filled with concrete. A travelling sixty-ton crane is purchased from Canada Foundry Company and installed in July. The two turbines are erected over the summer. Connections are made between the penstocks and turbines in September. Radial gates are installed in the intake dam in October while the generators are put in place. Plant is started up on November 21. Transmission lines are in place in early December. Power flows to Vancouver on December 17.

1912 Commercial generation begins in January when the second unit is ready. Contracts are signed to supply power to Bellingham and to Clayburn Brick, near Abbotsford.

1913 Contract is signed to supply power to BC Electric for twenty years. An agreement is made to purchase two more turbines from Escher-Wyss. Canadian General Electric is selected to supply the generators and transformers; Canadian Westinghouse to supply switchboard gear. Work on site, including raising the intake dam, done in preparation for additional units. Abernethy and Lougheed Logging Company purchase the right to cut forty million board feet of timber from areas to be flooded.

1914 Work suspended. Equipment already ordered put into storage.

1916 Unit 3 installed to meet contract obligations with BC Electric. Western Canada Power is reorganized, becomes Western Power Company of Canada.

1917 BC Electric and Western Power settle dispute over the failure of Western Canada Power to install a fourth unit on the schedule agreed to in 1913.

1920 BC Electric takes over Western Power. Robert F. Hayward leaves the company.

1921 Western Power closes its offices in Vancouver and office staff move to BC Electric offices. Major floods at Stave Falls in October. Planning and site preparation work begin for installing the fourth unit and raising the dams.

1922 Work done on raising the intake dam and raising and filling the sluice dam (which becomes the main dam). Work undertaken at Blind Slough in preparation for a new dam there. The contract to supply power to Bellingham is renewed for another ten years. Electrification of the railway begins. Work begins on overhaul of units 2 and 3. Unit 4 is put into operation in November. Survey work is begun for tunnel between Stave and Alouette Lakes.

1923 Decision is made to a install fifth unit in the existing powerhouse rather than on its own site below the main dam. Unit no. 1 is overhauled. West wing wall is raised. Work begins on replacing the Taintor gates in the intake dam with Stoney roller gates. The Taintor gates are moved to the new Blind Slough dam; 299 men are employed on construction work in September. Work begins on powerhouse extension for unit 5 in the fall. Drilling of Alouette tunnel is completed in December.

1924 Canadian Allis Chalmers given contract to supply fifth turbine. Steel shingle-bolt flume is built into the main dam. The Blind Slough dam is completed. Railway line is built to Alouette Lake. Survey work is carried out at Ruskin site. Ongoing problems with the operation of the new Stoney roller gates. Work begins on preparing the site for a new outdoor switching station, but subsequent slides make this site untenable and a decision is made to extend the tailrace deck and put the switchyard there. Decision is made to rebuild all the generators to 17,000 KVA. Work continues on powerhouse extension for unit 5.

1925 The tunnel between Stave and Alouette Lakes is completed in May. Unit 5 is started-up in September, with Premier John Oliver in attendance. Site clean-up begins with the dismantling of temporary canvas-roofed bunkhouses. The total cement used between 1921 and 1925 is 291,223 sacks.

1926 Alouette dam is completed.

1927 Work begins on the Alouette powerhouse.

1928 Alouette powerhouse is completed and in operation. Work begins at site of Ruskin plant.

1929 Contracts for Ruskin project awarded.

1930 Ruskin plant begins generating power in November. The grandson of John Work, Premier Simon Fraser Tolmie, attends the ceremony.

1932 Reservoir behind the Ruskin plant is named Hayward Lake in honour of Robert F. Hayward.

1935 "Silver frost" in the Fraser Valley downs powerlines from Stave Falls to the Fraser Valley and Vancouver.

1938 Second unit at Ruskin is installed and operating.

1944 Record low precipitation. Stave Falls Railway stops operating.

1949 Western Power Company of Canada is dissolved.

1950 Third unit at Ruskin comes onstream.

1959 BC Engineering Company completes its report, *Alouette-Stave-Ruskin: Preliminary Investigation of Proposals for Redevelopment as a Source of Peaking Capacity for the B.C. Electric System*, for BC Electric.

1962 The BC Hydro and Power Authority is created, combining the BC Power Commission and BC Electric.

1969 International Power and Engineering Consultants complete the *Alouette, Stave and Ruskin Power Development Study* for BC Hydro.

1981 BC Hydro's Hydroelectric Generation Projects Division begins studying redevelopment rehabilitation options at Stave Falls.

1983 Hayward Lake Recreation Area opens.

1984 Decision is made to rehabilitate Stave Falls plant. Last operational staff to live in the Stave Falls campsite leave.

1989 Decision to rehabilitate Stave Falls plant is reviewed.

1991 Recommendation is made to redevelop Stave Falls site.

1995 Provincial government grants BC Hydro energy project certificate for redevelopment project at Stave Falls.

1999 New powerhouse comes into service.

2000 The 1912 Stave Falls plant is shut down.

2001 The new visitor centre opens in the rehabilitated 1912 powerhouse.

Acknowledgements

THIS BOOK PROJECT, initiated by the Power Pioneers and BC Hydro, benefited from the help of many. Hugh Wilson, who started the project and drafted the first three chapters, kindly suggested to the editorial committee that I could finish the project. I am grateful to him, Colin Gurnsey, and the editorial committee, who, with the support of Drew Dunlop, gave me this opportunity. Hal Kalman of Commonwealth managed the project and wrote the "Stave Falls Railway" sidebar. Rob West, of Commonwealth, did research and wrote the "Technology at a Crossroad" sidebar.

A cross-section of people were interviewed: some grew up at Stave Falls, others worked there, and some have had a long association with the area or with BC Hydro. Some are quoted directly in the text; others provided background information. Many lent photographs. Thank you to Randy Bourne, John Boots, Roger Bruder, Warren Byrnell, Perry Cleven, Tony Dandurand, Bob Ellis, Frances Fleming, Doug Forster, Roy Forster, Jacqueline Fraser, Brenda Froese, Bob Heinrich, Howard Jackman, Russ Lamont, Flora McIntyre, Violet Middleton, Doug Noble, Philip Northrop, Paul Rapp, Ken Ruddick, Vera Ruddick, Jack Scott, Ralph Salsbury, Milton Sorley, Charmaine Schultz, Roy Taylor, Glen Truax, and Dorothy Winders. The interviews, consolidated by Andrew Wilson, were conducted primarily by Hugh Wilson and Jennifer McMahon. Christine Molle and Mary Lastoria provided transcription services.

This project was overseen by a supportive editorial committee who submitted to interviews, answered questions, did research, and furnished photographs. Committee members included Gerry Bramhill, John Carmichael, Bill Chatterton, Patricia Crawford, Brian Elder, Elan Falvai, Colin Gurnsey, Phil Horton, Charlie Scott, John Sled, and Don Swoboda. Other people associated with BC Hydro also helped, including Charlotte Bemister, Wayne Carlow, Rachel Kirby, Peter Partridge, and Rory Tennant.

The materials preserved in archives were essential. The BC Electric Railway Company's records, held at Special Collections, University of British Columbia Library, were especially important. The research carried out by Brian Elder in this collection provided key material. Other important sources include Stan Carter's memoirs at the Mission Community Archives, which the Carter family kindly gave permission to quote from. The help of Valerie Billesberger and her volunteer staff at the Mission Community Archives was outstanding. The memoirs of Frances Fleming, Bill Gosbee, Ove Sorley, and Gwen Smith also proved invaluable.

Others who helped included Valerie Patenaude, Maple Ridge Museum; Kelly Harms, Chilliwack Museum and Archives; Kelly Nolin, British Columbia Archives; Pat Molesky, Glenbow Archives; reference staff, Vancouver Public Library, History Division; Josee Vallerand, Archives of the Canadian Railway Museum; Carol Haber, City of Vancouver Archives; and Dave Smith, archivist at the Stó:lō Nation. Bob Turner, Carol Banman, D. Bruce Gleig, and Cheryle Macdonald also furnished information. Rob Keip at GVRD Parks kindly arranged for us to use a drawing by Brandon Gabriel.

I am grateful to two individuals: Fred Braches, of Whonnock, who helped me understand local settlement patterns and ferreted out photographs; and Tony Dandurand, of the Kwantlen First Nation, who furnished research materials, commented on drafts, and answered questions.

My apologies if I have overlooked anyone. I hope this book stimulates others to think and write about Stave Falls; I am sure there is more to say.

MEG STANLEY

Sources

Sources are provided here for direct quotations, including interviews not directly attributed in the text. **Bold face** numerals refer to this book's page numbers; the *italicized* words following them are the initial words of the quoted passages.

1 *[It] formed . . .* E.E. Carpenter, "The Water Developments of The Alouette-Stave-Ruskin Group of the British Columbia Electric Railway Company, Limited," *The Engineering Journal*, January 1927, p. 29.

2 *The whole . . .* R.F. Hayward, "The Stave Falls Power Development of Western Canada Power Company, Limited," *Transactions of the Canadian Society of Civil Engineers*, 29, pt. 2, 1915, p. 388.

Oxygen is . . . "Stave Lake Valley," *Daily Columbian* (New Westminster), May 9, 1891.

4 *Though the . . .* Hayward, p. 387.

Some [residents] . . . Interview of Clarence Cheer, Kwantlen First Nation, February 2, 1996.

7 *A well-defined . . .* Charles A. Miller, *Valley of the Stave* (Surrey, BC: Hancock House, 1981), p. 89.

The news . . . Quoted in Keith Thor Carlson, ed., *You Are Asked to Witness: The Stó:lō in Canada's Pacific Coast History* (Chilliwack, BC: Stó:lō Heritage Trust), p. 36.

Habitation in . . . Interview of Tony Dandurand, Kwantlen First Nation, by Hugh Wilson (H.W.), Fall 2000.

9 *In my last . . .* Quoted in B.A. McKelvie, *Fort Langley: Birthplace of British Columbia*, 2nd ed. (Victoria: Porcepic Books, 1991), pp. 55–56.

10 *Monday [September] 14th . . .* Reprinted with the permission of the publisher from Morag Maclachlan, ed., *The Fort Langley Journals, 1827–30*, (Vancouver: UBC Press, 1998), after p. 128. All rights reserved by the publisher.

11 *much wished . . .* Quoted in Kwantlen First Nation/Millennia Research Ltd., "The Stave Reservoir Archaeological Inventory and Impact Assessment Project," report prepared for BC Hydro, November 1997, p. 19.

13 *There was no . . .* Mission Community Archives (MCA), Foster and Sophie Carter Family fonds, Stan Carter, "Ruskin as I Knew It from 1890."

You may find . . . MCA, Foster and Sophie Carter Family fonds, Stan Carter's memoirs, p. 5.

14 *sheltered by . . .* J.A. Kirk, "Subdivision of Townships and Fixing Boundary of Railway Belt in British Columbia," in Canada, *Sessional Papers*, No. 14, Pt. 2 (Department of the Interior Annual Report) (Ottawa: Queen's Printer, 1890), p. 59.

passed from its . . . "Stave Lake Valley."

Stave River station . . . Quoted in Sheila Nickols et al., *Maple Ridge: A History of Settlement* (Maple Ridge, 1972), p. 52.

17 *The formation . . .* "Undertaking of Power Company is Highly Favoured by Nature," *Columbian* (New Westminster), August 13, 1910.

18 *It is conceded . . .* Quoted in Pierre Berton, *Niagara: A History of the Falls* (Toronto: McClelland and Stewart, 1992), p. 206.

20 *purchase, lease . . .* "Memorandum of Association of the British Columbia Timber Land and Electric Company, Limited Liability," *British Columbia Gazette*, February 2, 1893, p. 93.

Been maturing . . . British Columbia Archives (BCA), GR 1006, Water Rights Branch: Correspondence, 1889–1979, File 5, "Stave River File: Judgements, correspondence, etc.," James Leamy, crown timber agent, May 26, 1899, commenting on application by J.B. Ferguson for water power on Stave River in Twp. 15.

Take and use . . . British Columbia, An Act Respecting the incorporation of the Stave River Electric and Power Company, Limited Liability, *Statutes of British Columbia 1895*, chap. 67 (Victoria: Queen's Printer, 1895).

23 *put all their . . .* "Model Camp is at Stave Lake," *News-Advertiser* (Vancouver), February 20, 1911.

to use water . . . BCA, GR 1442, Order in-Council 225/00, file 13-14-5, "Stave Lake Power Co. Ltd . . ." MF Roll B4475, Box 4.

The Company have . . . Special Collections and University Archives, UBC Library. British Columbia Electric Railway Company fonds, 1896–1961 (BCER), Box 180, f. 11, Prospectus, October 1, 1900, p. 1.

The Company has . . . BCER, Box 180, f. 11, Prospectus, October 1, 1900, p. 4.

24 *In the spring of . . . Fraser Valley Record*, March 12, 1958, cited in MCA, clipping file no. 882.1.

25 *This year a . . .* BCER, Box 180, f. 11, "Letter to the Shareholders of Stave Lake Power Co.," November 10, 1906, p. 1.

26 *The action of . . .* "Work Suspended at Stave Lake," *Daily News* (New Westminster), October 23, 1907.

After careful . . . BCER, Box 180, letter from C.H. Cahan to Royal Securities Corp., May 12, 1909.

27 *August 17, 1908 . . .* BC Hydro, BC Electric Letterbook, 1909, "Stave Lake Power Company."

28 *February 11, 1910 . . .* BC Hydro, BC Electric Letterbook, 1910, "Stave Lake Power Company."

At the instance . . . BC Hydro, Stave Falls Records (SFR), Box A, letter to C.H. Cahan from R.F. Hayward, January 31, 1911.

29 *Selling power . . .* SFR, Box A, "Report on Construction, 30th June 1909–9th November 1911."

30 *I don't see . . .* BCER, Box 95, f. 12, letter from C.H. Cahan to R.F. Hayward, January 12, 1912.

He was reducing . . . BCER, Box 152, letter from G.R.G. Conway, chief engineer, BCER, to George Kidd, general manager, BCER, August 26, 1915.

certain factors . . . BC Hydro Records, letter from J.I. Newell, electrical superintendent, BCER, to George Kidd, General Manager, BCER, May 7, 1920, p. 3.

31 *The British Columbia . . .* George Kidd, "Company Takes Over Western Power Company," *BC Electric Employees Magazine* (BCEEM), January 1921, p. 8.

32 *by the construction . . .* BCER, Box 185, Western Canada Power Company, *Annual Statement*, 1911, p. 6.

33 *marvel of . . .* "Electrical Features of Alouette Plant," BCEEM, June 1927, p. 4.

of bending . . . British Columbia Power Corporation Limited, *Ruskin Power Development Completed November 1930* (pamphlet), pp. 3, 5.

35 *severe vibrations . . .* BC Hydro, Box FG 8041-01, letter from R.M. Skinner, superintendent of construction, to J.I. Newell, electrical superintendent, February 16, 1926.

36 *the work proved . . .* SFR, Box A, "Report on Construction, 30th June 1909–9th November 1911."

37 *The Marion steam . . .* Charles A. Miller, *Valley of the Stave* (Surrey, BC: Hancock House, 1981), p. 86.

Until the railway . . . SFR, Box A, Western Canada Power Company, "Construction Report," March 1, 1911.

38 *Thousands of feet* . . . Miller, p. 61.

Might it not . . . "The Crime of Negligence," *Courier* (New Westminster), October 7, 1911.

The arrangement . . . SFR, Box A, General Manager to C.H. Cahan, Royal Securities, November 22, 1910.

39 *In order to* . . . SFR, Box A, Western Canada Power Company, "Construction Report," March 1, 1911.

The cost of . . . SFR, Box A, "Report on Construction, 30th June 1909–9th November 1911."

40 *quite a pretentious* . . . "Undertaking of Power Company is Highly Favoured by Nature," *Columbian* (New Westminster), August 13, 1910.

A visitor . . . "Model Camp is at Stave Lake," *News-Advertiser* (Vancouver), February 20, 1911.

On proceeding further . . . Ibid.

41 *At night* . . . Ibid.

Accidents have been . . . "Undertaking of Power Company is Highly Favoured by Nature."

42 *Without the railway* . . . SFR, Box A, Western Canada Power Company, "Construction Report," March 1, 1911.

44 *As I told* . . . BCER, Western Canada Power (WCP), Letterbook No. 9 (LB), letter to Mr. McLeod, Star Realty, from R.F. Hayward, general manager, July 18, 1911 (letter no. 315).

46 *Owing to various* . . . BCER, WCP, LB 5, letter to A.R. Doble, secretary, Bank of Montreal, from R.F. Hayward, general manager, September 14, 1910 (letter no. 455).

We have received . . . BCER, WCP, LB 6, telegram to A.R. Doble, secretary, Bank of Montreal, from C.H. Cahan, September 26, 1910.

After all . . . BCER, WCP, LB 6, letter to James Maxie, superintendent of construction, from R.F. Hayward, general manager, November 23, 1910 (letter no. 419).

April 20, 1910 . . . BC Hydro, BC Electric Letterbook, 1910, "Stave Lake Power Company."

49 *There was sufficient* . . . BCER, Box 95, f. 13, letter to C.H. Cahan from R.F. Hayward, general manager, October 15, 1910.

The heavy rains . . . Ibid.

Yesterday, Friday . . . BCER, WCP, LB 4, letter to F.F. Busteed, general superintendent, CPR, from William McNeill, June 11, 1910.

A case marked . . . BCER, WCP, LB 9, letter to Messrs. McLennan, McFeely and Co., from Purchasing Agent, July 31, 1911 (letter no. 414).

52 *It was impossible* . . . "Fifth Unit at Stave Falls Now in Operation," *BCEEM*, October 1925, p. 9.

54 *Had this plant* . . . R.F. Hayward, "The Stave Falls Power Development of Western Canada Power Company, Limited," *Transactions of the Canadian Society of Civil Engineers*, 29, pt. 2 , 1915, p. 413.

58 *highly bureaucratic* . . . Robert A.J. McDonald, *Making Vancouver: 1863–1914* (Vancouver: UBC Press, 1996), p. 98.

enjoyed higher . . . Ibid., p. 101.

duty to look . . . Letter from R.M. Horne-Payne to J. Buntzen, March 21, 1900, quoted in Patricia Roy, "The British Columbia Electric Railway and Its Street Railway Employees: Paternalism in Labour Relations," *BC Studies*, 16 (Winter 1972–73), p. 3.

59 *lords and masters* . . . Interview of Charlie Scott by Meg Stanley (M.S.), 2000.

60 *breadwinners with wives* . . . Craig Heron, *The Canadian Labour Movement: A Short History* (Toronto: James Lorimer, 1996), p. 165.

61 *great reader* . . . "This Man is a Great Reader," *BCEEM*, May 1926.

63 *start exciters* . . . Interview of Roger Bruder by H.W., 2000.

determine the . . . IBEW, "Agreement between BC Hydro and Power Authority and Locals 213, 230, 821, 993 of the IBEW relating to Wages and Working Conditions, Effective 30 April 1963."

an employee . . . IBEW, "Agreement between BC Hydro and Power Authority and Local 258 of the IBEW relating to Wages and Working Conditions, Effective 1 April 1969."

65 *the worst job* . . . Bill Gosbee, "Memoirs," January 1968, unpublished.

isolate and . . . Interview of Charlie Scott by M.S., April 2000.

66 *There is a* . . . SFR, Anonymous poem found in Daily Log Books, 1930.

67 *it can be* . . . Mark Brett, "Frank Couldn't Escape the Lure of Stave Lake," *Intercom Magazine*, April 17, 1981, p. 3.

a born diplomat . . . "Elected Councillor," *BCEEM*, March 1931, p. 9.

68 *S. Carter* . . . BCER, Box 96, File 2: "Stave Falls Village," memo from E.L. Stenstrom to R.F. Hayward, October 1918.

69 *a trip* . . . "Fred Elwell, Retired," *BCEEM*, October 1940, p. 13.

A new engineer . . . Bill Gosbee

it wasn't . . . "Stave Falls Items," *BCEEM*, February 1927, p. 25.

70 *a sheet of* . . . BC Hydro, Box FG8042-01, letter to J.I. Newell from Stave Falls, August 11, 1924.

when you got . . . Interview of John Carmichael by H.W., 2000.

which was apparently . . . BC Hydro, Box FG8042-01, "Memorandum re: Flood at Stave Lake, October 28 to October 31, 1921."

71 *Mike "Dutch"* . . . "Workman Electrocuted by 12,000 Volt Wire," *Coquitlam Star*, April 17, 1912.

the principle . . . British Columbia, *Report of Committee of Investigation on Workmen's Compensation Laws* (Victoria: King's Printer, 1916), p. 1.

72 *A sad fatality* . . . *BCEEM*, January 1927, p. 26.

you just looked . . . "Upgrade Sparks Old Memories," *Access*, December 1993.

even with . . . Interview of Roy Taylor by H.W., 2000.

73 *Your work* . . . SFR, Box A, File. "Emergency Strike Operations, Notice to All Employees," September 27, 1943.

vulnerability of . . . BC Hydro, Box FG8041-01, memo from J.F. Fraser to Mr. Ingledow, February 19, 1941.

76 *typical bush* . . . Frances Fleming, "Memories of Stave Falls, BC," unpublished.

100 *the bake-shop* . . . "Stave Falls Items," *BCEEM*, September 1926, p. 19.

the company . . . BCER, Box 96, File 2: Stave Falls Village, Western Power Co. of Canada Ltd., "Regulations for Company's Dwellings at Stave Falls, R.F. Hayward, 1918–1922."

81 *down in the* . . . Interview of Flora McIntyre by Jennifer McMahon (J.M.), 2000.

82 *our company* . . . "Stave Falls Items," *BCEEM*, September 1926, p. 19.

1928 Pontiac . . . Interview of Doug Forster by H.W., 2000.

No employee . . . Charles A. Miller, *Valley of the Stave* (Surrey, BC: Hancock House, 1981), p. 61.

84 *Mr. Welsh* . . . Interview of Flora McIntyre by J.M., 2000.

87 *got pregnant* . . . Interview of Perry Cleven by J.M., 2000.

looked into this thing . . . BCER, Box FG 8042-01, letter from T.H. Welsh, powerhouse superintendent, to Mr. J.I. Newell, electrical superintendent, April 5, 1922.

something to eat . . . "Championship Tennis at Stave Falls," *BCEEM*, August 1927, p. 14.

88 *peppy pair* . . . "Stave Falls Items," *BCEEM*, January 1928, p. 19.

90 *depressed world* . . . Frances L. Fleming, "The Life and Times of a Woman who Took Charge," *Vancouver Sun*, October 20, 1999, p. A13.

91 *no one* . . . Ove Sorley, "Memoirs," unpublished, p. 88.

rent housing . . . IBEW, "Agreement between BC Hydro and Power Authority and Local 258 of the IBEW relating to Wages and Working Conditions, 1965."

93 *wilds of the* . . . "Stave Falls Items," *BCEEM*, December 1927, p. 29, "Stave Falls Items," *BCEEM*, December 1927, p. 30.

As I grew . . . Miller, p. 16.

95 *capitalism by . . .* Quoted in Hugh Wilson, *Gaslights to Gigawatts: A Human History BC Hydro and its Predecessors* (Vancouver: Hurricane Press, 1998), p. 134.

its 512,000 . . . Ibid., p. 111.

96 *obsolete and . . .* International Power and Engineering Consultants Limited, "Alouette, Stave, and Ruskin: Power Development Study." Report prepared for the British Columbia Hydro and Power Authority, August 1969, p. v.

dependable peaking . . . Ibid., p. v.

suitable for . . . Ibid,. p. vi–3.

the intensive . . . Ibid.

99 *BC Hydro's . . .* BC Hydro and Power Authority, *Annual Report–1971* (Victoria, 1971), p. 9.

is all more . . . SFR, Box A, "Report on Construction, 30th June 1909–9th November 1911."

immediate benefits . . . D.R. Bakewell, "Stave Lake Reservoir: An Overview of Debris Removal and Recreation Development," report prepared for BC Hydro and Power Authority, Properties Division, August 1980, p. 12.

101 *electrical utilities . . .* Mark Jaccard, John Nyboer, and Timo Makinen, "Managing Instead of Building: BC Hydro's Role in the 1990s," *BC Studies*, Autumn/Winter 1991–92, nos. 91–92, p. 126.

aims at . . . Ibid., p. 113.

provide significantly . . . BC Hydro Hydroelectric Engineering Division, "Stave Falls Project Redevelopment: Feasibility Study," December 1991, p. v.

102 *inefficient (with only . . .* Ibid.

103 *a new intake . . .* Randy Bourne and Dennis Moore, "Stave Falls Powerplant Replacement: Offering Multiple Solutions," *Hydro Review*, December 1999, p. SR2.

PHOTO CREDITS

ii "Progress Report: Stave Falls Water Power Development, April 1925," courtesy of Kenneth and Vera Ruddick.

vii Courtesy of Doug Noble and family.

viii BC Hydro Information Centre (IC) B1057/188.

2 IC B1050/1

3 Map adapted from UBC Special Collections and University Archives, UBC Library, British Columbia Electric Railway Fond (BCER), Box 183, "The Western Canada Power Company, Limited," *Western Canada Power Annual Report, 1912.*

4 Courtesy of Flora McIntyre.

5 Courtesy of Doug Noble and family.

6 Greater Vancouver Regional District Parks and Brandon Gabriel.

7 British Columbia Archives (BCA) F-08505.

8 Maple Ridge Museum (MRM) P00085.

9 *Left:* BCA D-09086; *right:* BCA, PDP-00040.

10 BCA A-04313.

11 Archives of Canadian Railroad Historical Association.

12 Courtesy of Doug Noble and family.

13 Courtesy of Brian Byrnes, Whonnock.

14 Courtesy Mission Community Archives (MCA) Pr 27-1.

15 MRM P03350.

16 BCA I-52069.

19 City of Vancouver Archives, LGN 1017.

21 Map adapted from MCA, "Municipality of Mission, *c.* 1904," *inset:* adapted from BC Hydro, Stave Falls Records (SFR), "Western Canada Power, General Plan of Hydraulic Development Stave River Falls, 1910."

22 *Left:* Glenbow Archives NA 4428-1; *centre:* Association of Professional Engineers of the Province of British Columbia, *Yearbook, 1930; right:* IC, *BC Electric Employees' Magazine*, May 1924.

23 BCER.

25 Sholefield and Howay, *British Columbia from the Earliest Times to the Present*, vol. 3.

26 *Who's Who and Why: A Biographical Dictionary of Notable Men and Women of Canada and Newfoundland,* vol. 5.

28 Courtesy of Doug Noble and family.

29 MCA.

31 Courtesy of Doug Noble and family.

32 *Left:* IC; *centre:* BC Power Commission, "Ruskin Power Development Completed November 1930," courtesy of Miller family.

33 BC Hydro, Stave Falls Records (SFR).

34 "Report: Stave Falls Water Power Development, April 1925," courtesy of Kenneth and Vera Ruddick.

36 Courtesy of Miller family

37 IC B1094/167.

38 IC, Hayward Album.

39 IC B-1092/12.

40 MCA 123-26A.

41 *Left:* IC B1092/97; *right:* IC B1092/96.

42 Courtesy of Miller family.

43 Courtesy of Doug Noble and family.

44 IC B1093/82.

45 IC B1094/195.

46 IC B1093/33.

47 SFR.

48 "Report: Stave Falls Water Power Development, April 1925," courtesy of Kenneth and Vera Ruddick.

50 SFR, "Progress Report: Stave Falls Water Power Development, March 1923."

51 IC B1056/7

53 IC B1094/4

54 IC B1092/74.

56 Courtesy of Doug Noble and family.

58 Courtesy of Doug Forster.

59 Courtesy of Doug Noble and family.

60 IC B1054/2.

61 IC: *BC Electric Employees' Magazine*, May 1925.

62 Courtesy of Doug Noble and family.

63 MCA Pr 299-15.

64 IC: *Intercom*, April 1969.

65 Commonwealth Historic Resource Management Limited (CHRML).

66 Courtesy of Kenneth and Vera Ruddick.

67 Courtesy of Flora McIntyre.

68 MCA Pr 304-80.

70 MCA Pr 316-72.

71 Courtesy of Kenneth and Vera Ruddick.

73 Courtesy of Doug Noble and family.

74 Courtesy of Dorothy and Len Winders.

76 SFR.

77 MCA Pr 27-34.

78 *Left:* MRM P02559; *right:* Courtesy of Dorothy and Len Winders.

79 SFR, "Progress Report: Stave Falls Water Power Development, April–June 1923."

80 Courtesy of Dorothy and Len Winders.

81 Courtesy of Kenneth and Vera Ruddick.

82 Courtesy of Ralph Salsbury and family.

83 Courtesy of Doug Forster.

84 Courtesy of Ralph Salsbury and family.

85 *Left:* MCA Pr 172-4; *right:* courtesy of Flora McIntyre.

86 Courtesy of Kenneth and Vera Ruddick.

87 MCA Pr 153-27.

88 Courtesy of Doug Noble and family.

89 Courtesy of Flora McIntyre.

90 Courtesy of Flora McIntyre.

91 Courtesy of Kenneth and Vera Ruddick.

92 Courtesy of Flora McIntyre.

93 *Left:* courtesy of Miller family; *centre, right:* courtesy of Frances Fleming.

94 BC Hydro Stave Falls Powerplant Replacement Project Photos (PRP).

96 IC/ed white photographics.

97 IC/ed white photographics.

98 IC, GS 116.

100 IC B1084/4.

101 MCA Pr 306-58a.

102 PRP.

103 SFR/Paramount Pictures Corp. (Canada) Inc.

104 PRP.

105 PRP.

106 PRP.

107 BC Hydro MultiMedia Studio (MMS).

108 BC Hydro Advertising.

109 MMS.

111 CHRML.

Index

Italicized page references indicate photo captions.